Current Topics in Microbiology and Immunology

220

Springer
Berlin
Heidelberg
New York
Barcelona
Budapest
Hong Kong
London
Milan
Paris
Santa Clara
Singapore
Tokyo

Chromosomal Translocations and Oncogenic Transcription Factors

Edited by F.J. Rauscher, III and P.K. Vogt

With 28 Figures

Springer

Frank J. Rauscher, III, Ph.D.
Associate Professor
Chair, Molecular Genetics Program
The Wistar Institute
3601 Spruce Street
Philadelphia, PA 19104-4268
USA

Peter K. Vogt, Ph.D.
Division of Oncovirology, BCC 239
Department of Molecular and
Experimental Medicine
The Scripps Research Institute
10550 North Torrey Pines Road
La Jolla, CA 92037
USA

Cover illustration: *The cover depicts the histopathology and genetic pathology underlying alveolar rhabdomyosarcoma (ARMS). Background*: *A section of the ARMS tumor shows the small round cells which proliferate in sheets along a fibrous stroma giving rise to the characteristic "alveolar" phenotype. Inset*: *The transcription factor-encoding genes PAX3 and FKHR are translocated and fused as a result of the t(2;13) in ARMS. The PAX3-FKHR fusion oncogene is a powerful activator of transcription. See the article by F. Barr in this volume who kindly provided the tumor section.*

Cover design: Design & Production GmbH, Heidelberg

ISSN 0070-217X
ISBN-13: 978-3-642-64424-5 e-ISBN-13: 978-3-642-60479-9
DOI: 10.1007/978-3-642-60479-9

Softcover reprint of the hardcover 1st edition 1997
Library of Congress Catalog Card Number 15-12910

Typesetting: Scientific Publishing Services (P) Ltd, Madras

SPIN: 10537538 27/3020/SPS – 5 4 3 2 1 0 – Printed on acid-free paper

Preface

Transcriptional regulation of gene expression lies at the heart of almost every fundamental homeostatic process in biology including regulation of DNA synthesis, cell division, cellular differentiation, control of apoptosis, organismal development and organogenesis. The primary mediators at this level of gene regulation are transcription factors; nuclear proteins that are endowed with both sequence-specific DNA recognition and the ability to regulate transcription of the bound target gene. It is well established that tissue-specific transcription factors can initiate and maintain entire programs of cellular differentiation via proper regulation of the set of target genes that contain the cognate DNA recognition sequences. Essentially, the population of transcription factors which are present and active in a cell nucleus at a given time specifies the cellular phenotype. Thus, it is not surprising that deregulation of transcription factor function is a primary mechanism underlying neoplastic transformation and malignant progression. Many oncogenes and tumor-suppressor genes encode transcriptional regulators. Alteration of transcription factor function has the potential to generate novel cellular phenotypes, including phenotypes with frank oncogenic properties.

This volume is dedicated to the molecular and biochemical mechanisms of transcription factor dysfunction that lead to cell-type and organ-specific loss of growth control and subsequent malignant progression. We have chosen to focus on transcription factors that are activated as apparently dominant oncogenes, originating from disease-specific chromosomal rearrangements and translocations. Our interest is to explore the biochemical functions that are targets for mutational alteration in transcription factors. These include alteration of (i) subcellular or subnuclear localization, (ii) DNA binding affinity or selectivity, (iii) protein-protein interaction via hetero- or homo-oligomerization, and (iv) transcriptional regulatory potency. Almost every class of DNA binding domain, including ETS domains, homeobox, basic-helix-loop-helix, and zinc-fingers are represented among the cancer-specific disrupted transcription factors. A common theme involves fusion of the sequence-specific DNA recognition module to novel effector domains that serve to deregulate the fused transcription factor and result in oncogenesis. Transcription factors are commonly mutated in human oncogenesis, and this volume deals with numerous tumor types, including myeloid and lymphoid leukemia, both pediatric and adult, and the spectrum of soft tissue sarcomas, including Ewing's sarcoma, peripheral neuroectodermal tumors and rhabdomyosarcoma.

The volume opens with a chapter by P.M. Waring and M.L. Cleary on the homolog of *Drosophila* trithorax located at 11q23 in the human genome and involved in translocations to diverse other genes. As many as ten of these translocation partners of the trithorax homolog have now been characterized. Translocations involving 11q23 typically occur in acute lymphoblastic leukemias and subsets of acute myeloid leukemias. The common feature of the 11q23 translocations is a truncation of the trithorax homolog. In childhood B-cell acute lymphoblastic leukemia, a common translocation involves the transcription factor E2A that can be fused to the homeobox protein (Pbx) or to a member of the b-ZIP super-family known as hepatic leukemia factor (HLF). These translocations are reviewed in the chapters by M.P. Kamps and A.T. Look. An interesting property of the E2A fusion products is their ability to induce oncogenic transformation in experimental systems, a clear demonstration of their status as dominantly acting oncoproteins. The chapter by R. Baer, L.-Y. Hwang and R.O. Bash is devoted to the interaction of basic helix-loop-helix proteins and LIM domain proteins. Ectopic expression and activation of these transcription factors plays an important role in the development of T cell acute lymphoblastic leukemia. T.R. Golub, G.F. Barker, K. Stegmaier and D.G. Gilliland contribute a chapter on the *TEL* gene. TEL is another transcription factor disrupted in leukemias and able to fuse with an amazing variety of partners. It contributes DNA binding or HLA domains to fusion products that are important in myeloid and lymphoid leukemias in pediatric as well as adult populations. One of the now classical translocations involving two regulatory proteins is associated with acute promyelocytic leukemia. It is reviewed in the chapter by D. Grimwade and E. Solomon. This translocation fuses the retinoic acid receptor-α, a member of the steroid receptor family, with a novel gene, *PML*. The fusion product retains the ligand binding domain of the retinoic acid receptor. The wild-type *PML* gene is intriguing because it shows growth attenuating properties. The chapter by F.G. Barr then turns to solid tumors, reviewing the fusion of paired box and fork head family genes in alveolar rhabdomyosarcoma. The transcription factor generated by this translocation carries the DNA binding domain derived from the paired box progenitor and the transactivation domain of the fork head protein. The chapter by D. Ron brings a new molecular function into focus, a protein, TLS, that binds RNA instead of DNA. TLS is fused to the CHOP gene to produce a potent transcriptional activator important in myxoid liposarcoma. TLS is closely related to the EWS gene reviewed in the chapter by W.A. May and C.T. Denny. EWS also contains an RNA recognition motif, and in Ewing's sarcoma it is fused to the ETS family transcription factor FLI-1 or less frequently to other ETS genes such as ERG or ETV-1. The final chapter of the volume by F.J. Rauscher is devoted to the fusion of EWS1 and WT1 in small round cell tumors.

A major goal in gathering together these tales of transcription factor dysfunction in diverse neoplastic processes is to address the following questions: First, what are the molecular aspects of specificity which allow nonrandom alteration of a single transcription factor to lead to a highly specific disease process? Second, can these tumor-type specific transcription factor alterations be utilized for molecular diagnosis, prognosis or as disease-specific targets for cancer therapy? Designing small molecule or gene-based therapies that target the cellular oncogenic transcription factor is an attractive and potentially highly selective method of intervention. The models of altered transcription factor function collected in this volume will likely prove a fertile field for testing these concepts in the years to come.

We would like to thank the authors who have spent their valuable time in contributing to this volume. Their cooperation and expertise was crucial in obtaining a comprehensive, state-of-the-art synopsis of a complex topic.

Philadelphia, PA FRANK J. RAUSCHER, III
La Jolla, CA PETER K. VOGT

List of Contents

List of Contributors

(Their addresses can be found at the beginning of their respective chapters.)

Disruption of a Homolog of Trithorax by 11q23 Translocations: Leukemogenic and Transcriptional Implications

P.M. WARING and M.L. CLEARY

1 Introduction

Based on cytogenetic studies, chromosome band 11q23 was predicted to contain an important gene, disruption of which was suspected to play a crucial role in leukemogenesis. Within the past 3 years, one of these disrupted genes, *HRX* (also known as *Htrx1*, *MLL* or *ALL-1*), and ten of its fusion partner genes have been cloned. Insights into the physiological role of the HRX protein and the leukemogenic consequences of its disruption are beginning to emerge from the observations that HRX is a homologue of the *Drosophila* homeotic regulator trithorax and may exert its effects through recognition and modification of chromatin structure.

Department of Pathology, School of Medicine, Stanford University, Stanford, CA 94305-5324, USA

2 The Identification, Cloning and Characterization of *HRX*

2.1 HRX Is Identified by Its Involvement in 11q23-Associated Leukemias

Chromosome band 11q23 is a recurring site for cytogenetic alterations, mostly reciprocal translocations, in acute lymphoblastic leukemias (ALLs) and subsets of acute myeloid leukemias (AMLs). The physical location of the 11q23 leukemia locus was determined by fluorescence in situ hybridization (FISH) using a yeast artificial chromosome yB22B2 that spanned the *HRX* breakpoint (Rowley et al. 1990) (Fig.1). A candidate gene on chromosome 11 was designated *MLL* (mixed lineage leukemia or myeloid/lymphoid leukemia) (Ziemin-van Der Poel et al. 1991) and *ALL-1* (acute lymphocytic leukemia-1) (Cimino et al. 1991). The gene disrupted by t(4;11)(q21;q23) (Djabali et al. 1992; Gu et al. 1992b; Domer et al. 1993) and by t(11;19)(q23;p13.3) (Tkachuk et al. 1992) was cloned and the full length cDNA sequence revealed a single long open reading frame of 11 904 nucleotides, encoding a protein that was named HRX (homologue of trithorax) (Tkachuk et al. 1992) and Htrx1 (human trithorax 1) (Djabali et al. 1992) in view

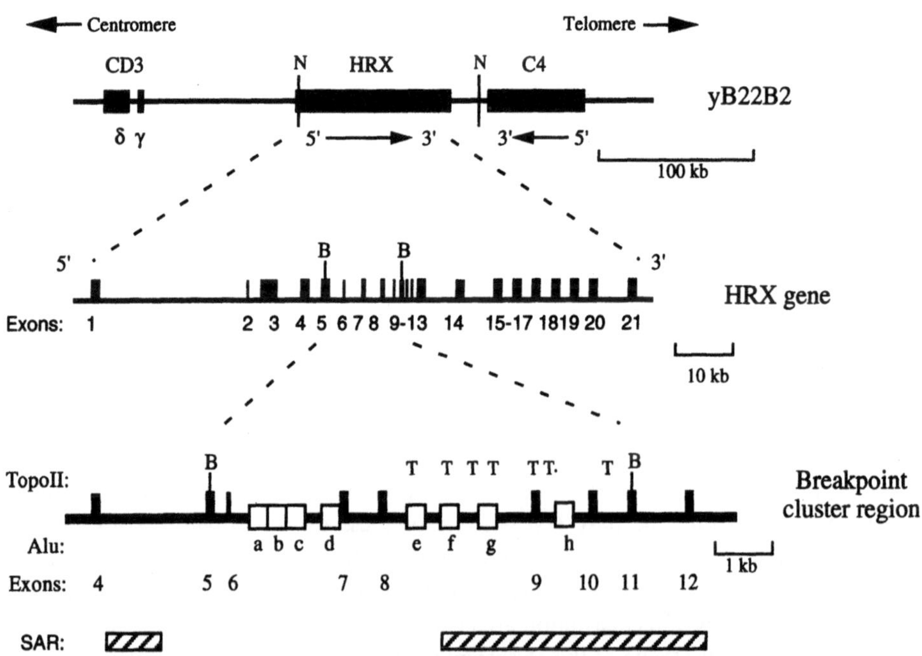

Fig. 1. The structure of the *HRX*-containing yeast artifical chromosome (yB22BS), the *HRX* gene, and *HRX* breakpoint cluster region. *Black vertical bars* represent exons 1–21; *horizontal black bars* represent scaffold attachment regions (SAR). *B*, *Bam*H1 site, defines the 8.3 kb breakpoint cluster region covered by the 0.84 kb cDNA probe; T, high stringency topoisomerase II consensus sites

of its partial sequence homology to the *Drosophila melanogaster* protein trithorax (trx). Due to the functional significance of this homology we prefer the term HRX, usage of which is not intended to diminish the contributions of others.

2.2 Structure of the *HRX* Gene

The *HRX* gene is composed of at least 21 exons spanning approximately 100 kb of genomic DNA (ZIEMIN-VAN DER POEL et al. 1991; GU et al. 1992b; TKACHUK et al. 1992) (Fig. 1). Several investigators have shown that almost all *HRX* breakpoints are tightly clustered in an 8.3 kb genomic region containing *HRX* exons 5–11 (CHEN et al. 1991; GU et al. 1992b; TKACHUK et al. 1992; DJABALI et al. 1992; MORGAN et al. 1992; CORRAL et al. 1993; HILDEN et al. 1993; IIDA et al. 1993a; DOMER et al. 1995; BROEKER et al. 1996). The entire breakpoint cluster region has been sequenced (GU et al. 1994) and found to contain eight *Alu* repeat elements in the same transcriptional orientation as *HRX* (DJABALI et al. 1992; GU et al. 1994), several topoisomerase II (topo II) recognition consensus sites (DOMER et al. 1993, 1995; NEGRINI et al. 1993; GU et al. 1994; BROEKER et al. 1996), and a high affinity scaffold attachment region (SAR) (BROEKER et al. 1996) (Fig. 1).

2.3 Predicted Structure of the HRX Protein

The cloned *HRX* cDNA predicts a polypeptide of 3969 amino acids comprising an unusually large 431 kDa protein (TKACHUK et al. 1992) that is highly basic (pI = 9.70) and rich in serine (13.8%) and proline (8.9%) residues. The murine homologue of HRX, mouse ALL-1, shows 90.8% overall identity with HRX and shares all of its structural motifs (MA et al. 1993). Within its NH_2-terminal third, HRX shows no homology to *Drosophila* trithorax but contains several distinctive motifs (Fig. 2) including: a triplet of Arg-Gly-Arg-Pro sequences that constitute three AT hook motifs; an acidic-basic repeat reminiscent of those found in RNA binding proteins, e.g., 70 kDa U1 small nuclear ribonucleoprotein (snRNP) (THEISSEN et al. 1986) and the human RD protein (SUROWY et al. 1990); a cysteine-rich region with homology to DNA methyltransferases (MTases); and a lysine-rich region with homology to the COOH-terminal tail of late histone H1 (TKACHUK et al. 1992; GU et al. 1992b; DOMER et al. 1993; PARRY et al. 1993, PMW, unpublished observations). The central portions of HRX, murine ALL-1, and trx contain two regions that encompass several unusual zinc-finger-like domains (MAZO et al. 1990; MA et al. 1993; TKACHUK et al. 1992; GU et al. 1992b; PARRY et al. 1993) which can be configured into the Cys_4-His-Cys_3 pattern of PHD fingers that share some features with the LIM and RING fingers that mediate protein-protein interactions (AASLAND et al. 1995). These structures are present in a diverse group of proteins (AASLAND et al. 1995) and are likely to be functionally important since deletion of

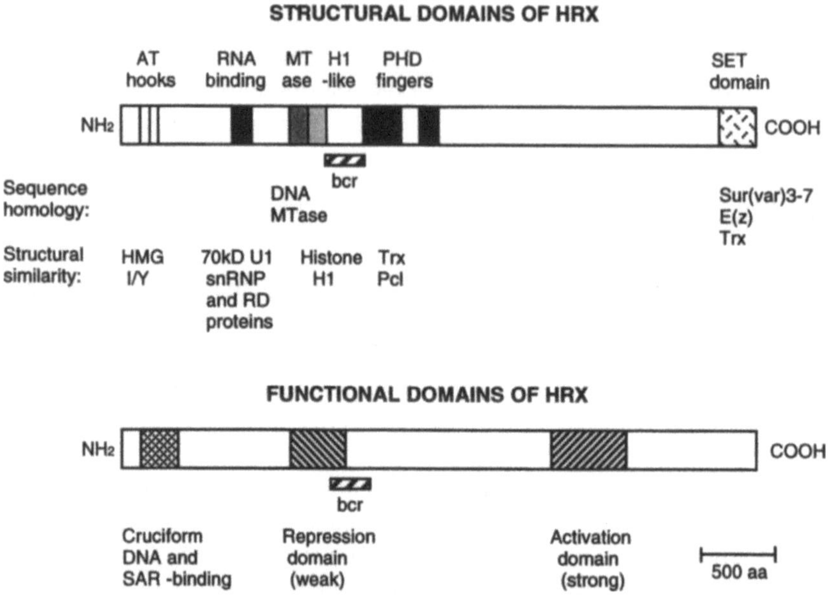

Fig. 2. The structural and functional domains of HRX

one of the zinc finger domains of trx was lethal to *Drosophila* (Mazo et al. 1990). The COOH-terminal 215 residues of HRX show 64% identity with the comparable, functionally uncharacterized, region of trithorax. The homology is highest (75% identity) in the last 132 amino acid segment, a region that also shows a high degree of homology with the COOH-terminals of the *Drosophila* proteins encoded by the enhancer of zeste (*E*(*z*)) (40% identity) (Jones and Gelbart 1993), the suppressor of position-effect variegation (PEV) gene *Su(var)3–9* (34% identity) (Tschiersch et al. 1994), the G9a gene located within the human MHC locus (35% identity) (Milner and Campbell 1993), and an uncharacterized yeast sequence, YHR9 (50% identity) (Xiangmin Cui and MLC, unpublished observation). The region, termed a SET domain (suppressor of variegation, enhancer of zeste and trithorax) (Tschiersch et al. 1994), has been conserved over great evolutionary distance suggesting an important functional role.

Like HRX, *Drosophila* trx is also an unusually large protein, comprising 3759 residues, and is predicted to exist in two isoforms (368 kDa and 404 kDa) (Mazo et al. 1990; Sedkov et al. 1994; Kuzin et al. 1994). Notably, none of the NH_2-terminal motifs of HRX are conserved in the fly protein. AT-hook motifs are present in several DNA-binding proteins, including the high mobility group (HMG)-I(Y) of nonhistone chromosome -binding proteins that bind to AT-rich sequences within the minor groove of DNA (Reeves and Nissen 1990). Nearby are two SPKK motifs belonging to the "Ser-Pro-basic-basic" consensus that are repeated several times in the tails of histones H1 and H2B which also bind to the minor groove of

DNA and are important for chromatin condensation (SUSUKI 1989; CHURCHILL and SUSUKI 1989). AT hook and SPKK motifs are both substrates for cdc2 kinase phosphorylation (WOLFE 1991; REEVES 1992) suggesting that HRX could be differentially phosphorylated during the cell cycle. The cysteine-rich region of HRX and murine ALL-1 with similarity to mammalian DNA MTase (MA et al. 1993; DOMER et al. 1993) is a zinc-binding domain which comprises a portion of MTase that is partly responsible for discriminating methylated and unmethylated DNA (BESTOR et al. 1988; BESTOR 1992). Curiously, the region of HRX with homology to the 70 kDa U1 snRNP is also shared with MTase (LEONHARDT et al. 1992). These features suggest that the NH_2-terminal region of HRX may recognize methylation status among other structural features of mammalian DNA. Notably, *Drosophila* DNA is unmethylated and these regions are absent from trx.

2.4 *HRX* Expression

Northern blot analyses indicate that *HRX* is widely, perhaps ubiquitously, expressed. Low abundance *HRX* mRNAs were detected in T and B lymphocytes, myeloid, erythroid, epithelial, fibroblast, hepatic and glial cell lines (ZIEMIN-VAN DER POEL et al. 1991; CIMINO et al. 1991, 1992; GU et al. 1992a, 1992b; TKACHUK et al. 1992; DJABALI et al. 1992; PARRY et al. 1993; MCCABE et al. 1992; YAMAMOTO et al. 1993a). Like *trx* (MAZO et al. 1990; SEDKOV et al. 1994; KUZIN et al. 1994), *HRX* is usually expressed as two differentially spliced transcripts of 15 and 13 kb (ZIEMIN-VAN DER POEL et al. 1991; CIMINO et al. 1991, 1992; GU et al. 1992a, 1992b; TKACHUK et al. 1992; YAMAMOTO et al. 1993a; MCCABE et al. 1994), the smaller being the result of alternate splicing which removes a region containing the AT hook motifs (DOMER et al. 1993). *Trx* has up to five alternately spliced transcripts (11–14 kb) which have different developmental profiles (BREEN and HARTE 1991, 1993).

2.5 HRX Is a Nonclassical Transcription Factor

Like HMG proteins, HRX appears to recognize the structure rather than the sequence of its target DNA, suggesting that it is not a classical sequence-specific transcription factor. The AT hook region of HRX was shown to bind synthetic cruciform and SAR DNA in gel mobility shift assays (ZELEZINK-LE et al. 1994; BROEKER et al. 1995) suggesting that this portion of HRX is responsible for the recognition of bent DNA, e.g., in AT-rich regions. The functional significance of these interactions in vivo is, however, unclear. Functional transcriptional regulatory domains have been identified (Fig. 2) and a region of HRX distal to the leukemia fusion sites was found to be a potent transcriptional activator whereas a region proximal to the fusion sites was shown to be a weak repressor of transcription (ZELEZNIK-LE et al. 1994).

2.6 Trx and HRX Are Both Positive Regulators of Homeotic Gene Expression

Trithorax was the first known member of a group of approximately 40 *trans*-acting factors known as the trithorax group (*Trx*-G) which, together with the opposing *Polycomb* group (*Pc*-G), maintain the expression of homeotic (*HOM-C*) genes in *Drosophila* (*Paro* 1990, 1993; KENNISON 1993) by interacting with *cis* regulatory elements in their promoters (CASTELLI-GAIR and GARCÍA-BELLIDO 1990; ZINK et al. 1991; KUZIN et al. 1994; CHAN et al. 1994; GINDHART and KAUFMAN 1995). In *Drosophila*, the homeotic genes are clustered in two large loci, the *Antennapedia* (*ANT*-C) and *bithorax* (*BX*-C) complexes, which are differentially expressed along the anterior-posterior axis of the developing embryo and are responsible for determining the structures and appendages that are characteristic for each body segment of the fly (McGINNIS and KRUMLAUF 1992). Complex genetic analyses in *Drosophila* demonstrated that members of the *Trx*-G and *Pc*-G are necessary for the maintenance, but not initiation, of homeotic gene expression patterns that are predetermined by the transiently expressed gap and segmentation genes (PARO 1990; KENNISON 1993). *Pc*-G gene products are required to maintain a repressed transcriptional state whereas *Trx*-G gene products ensure continued homeotic gene expression. Members of each group therefore appear to serve a sort of epigenetic memory function that fixes embryonic cell fate by imprinting a determined transcriptional state on their subordinate *Hox* genes. How this is achieved is unclear, but may reside in the ability of *Trx*-G and *Pc*-G gene products to somehow sense whether the chromatin conformation around the regulatory regions of their target gene is in a transcriptionally active "open" or an inactive "closed" form at the time of gap and segmentation gene activity and to fix that state so that it is inherited in a clonal manner throughout development (PARO et al. 1993). Such proteins are known to form large multiprotein complexes that colocalize on polytene chromosomes at promoter sites of *ANT*-C, *BX*-C, and other loci (PARO 1993; KUZIN et al. 1994); however, the molecular mechanisms of how the various *Pc*-G and *Trx*-G proteins interact and recognize and determine chromatin structure in the vicinity of the regulatory regions of *Hox* genes are unknown.

Mutations in the fly *Trx* gene result in anterior homeotic transformations of thoracic and abdominal segments like those seen in sex comb reduced (*Scr*), *Antp, Ubx, abd-A* and *Abd-B* mutants (BREEN and HARTE 1991), suggesting that *Trx* is required for the proper restricted expression of these genes. As in *Drosophila*, tight regulation of mammalian *Hox* expression boundaries and dosage are also critical in determining segment identity since altered expression of various *Hox* genes in mouse embryos result in homeotic transformations (McGINNIS and KRUMLAUF 1992; KRUMLAUF 1994). HRX-deficient mice generated by targeted disruption of *HRX* in embryonic stem cells also displayed a homeotic phenotype (YU et al. 1995). Homozygous HRX-deficiency is embryonic lethal and is accompanied by abolition of *Hox* expression.

Heterozygous HRX-deficient mice are growth retarded, display hematopoietic abnormalities such as anemia and B cell aberrations, and exhibit axial skeletal defects with anterior homeotic transformation manifest as loss of the most posterior pairs of ribs and posterior displacement of the anterior boundaries of *Hox* expression patterns. This contrasts with the posterior homeotic transformation of vertebral identity observed in *bmi-1*-deficient mice (VAN DER LUGT et al. 1994). Interestingly, the opposite phenotype, namely an anterior transformation, was observed in Eμ-*bmi-1* transgenic mice (ALKEMA et al. 1995), illustrating the importance of dosage and restricted *Hox* expression patterns. Since *HRX* and *bmi-1* are, respectively, human homologues of *Trx* (TKACHUK et al. 1992) and posterior sex combs (*Psc*, a member of the *Pc*-G) (BRUNK et al. 1991; VAN LOHUIZEN et al. 1991), these observations provide strong evidence that *HRX*, like *trx*, is a positive regulator of *Hox* genes and that the opposing regulatory pathways that control *Hox* expression and function in *Drosophila* have been conserved in vertebrates.

3 The Role of HRX in Leukemogenesis

3.1 HRX Involvement in De Novo ALL and AML

The *HRX* gene has been shown to be rearranged by 11q23 abnormalities whose reciprocal partners are located at almost 30 cytogenetically diverse loci including 1p32, 1q32, 1q21, 2p21, 3q23–25, 4q21, 5q31, 6p12, 6q27, 7p15, 7p22, 9p11, 9p22, 10p11–13, 10q22, 11q13, 12p13, 15q15, 16p13, 17q21, 17q25, 18q21, 19p13.1, 19p13.3, 22q12, and Xq13 (GU et al. 1992b; KEARNEY et al. 1992; TKACHUK et al. 1992; CORRAL et al. 1993; HUNGER et al. 1993; LIDA et al. 1993b; JANI SAIT et al. 1993; KOBAYASHI et al. 1993; MORRISEY et al. 1993; NAKAMURA et al. 1993; PRASAD et al. 1993, 1994; THIRMAN et al. 1993a, 1994; BERNARD et al. 1994; HEIGHT et al. 1994; LEBLANC et al. 1994; MCCABE et al. 1994; PARRY et al. 1994; RUBNITZ et al. 1994a; SORENSEN et al. 1994; CHAPLIN et al. 1995a, b; FELIX et al. 1995; HERNANDEZ et al. 1995; MITANI et al. 1995; TSE et al. 1995). Notably, this degree of promiscuity is unprecedented among nonrecombinatorial genes and suggests that the breakpoint cluster region of *HRX* may be a recombination "hot spot" harboring highly recombinogenic sequences or structures.

Reciprocal translocations disrupting *HRX* account for the majority of 11q23 abnormalities seen in early B cell precursor ALL and myelomonocytic (FAB-M4) and monocytic (FAB-M5) AMLs (BOWER et al. 1994a; CHEN et al. 1993b; CIMINO et al. 1993, 1995a,b; PUI et al. 1994; RUBNITZ et al. 1994a; SORENSEN et al. 1994; STOCK et al. 1994). That HRX-associated leukemias can be of either lymphoid or myeloid lineage or express markers of both suggests that HRX mutations occur in hematopoietic stem or early progenitor cells. Although uncommon in adults (STOCK et al. 1994; BOWER et al. 1994a; CIMINO et al. 1995a), leukemias bearing *HRX* rearrangements account for 60%–80% of all infant leukemias (CHEN et al. 1993b; CIMINO et al. 1993, 1995b; RUBNITZ et al. 1994a; PUI et al. 1994; SORENSEN et al. 1994; Cimino et al 1995a; MARTINEZ-CLIMENT et al. 1995). Studies of infant twins with concordant leukemia have shown that translocations disrupting *HRX* can occur in utero (FORD et al. 1993; GILL SUPER et al. 1994). The short latency in such cases suggests a recessive mechanism, however, whether mutation of *HRX* alone is sufficient for the development of the leukemia or instead renders the cells highly susceptible to further leukemogenic events remains unknown.

3.2 Partial Duplication of *HRX* Is Common in Leukemias Without Cytogenetic Abnormalities of 11q23

Several adult AMLs without obvious karyotypic abnormalities or with trisomy 11 as the sole cytogenetic abnormality have now been described with partial tandem internal duplications of *HRX* in which the duplicated NH_2-terminal portion of HRX is fused in frame with itself just upstream of exon 2 (SCHICHMAN et al. 1994a, b, 1995). Such cases may represent a special subgroup since many of these leukemias were of myeloblastic (FAB-M1) or myelocytic (FAB-M2) subtype (SCHICHMAN et al. 1995) and AMLs with trisomy 11 characteristically have a stem cell phenotype (SLOVAK et al. 1995). Self-fusion potentially represents a new mechanism of leukemogenesis and indicates that mutation of HRX may be leukemogenic without a contribution from a heterologous fusion partner (SCHICHMAN et al. 1995).

3.3 *HRX* Involvement in Therapy-Related AML

HRX rearrangements are present in approximately 80% of therapy-related AMLs (t-AMLs) that arise following chemotherapy with epipodophyllotoxin derivative drugs (e.g., etoposide, tenoposide) for a primary unrelated malignancy, but, notably, are not present in t-AMLs that arise in patients treated with other agents (HUNGER et al. 1993; GILL SUPER et al. 1993; FELIX et al. 1993, 1995; LEBLANC et al. 1994; BOWER et al. 1994b; DOMER et al. 1995). Since the epipodophyllotoxins target DNA-topoisomerase II and the breakpoint cluster region of *HRX* contains several topo II consensus binding sites, there is the intriguing possibility that this region may be readily subjected to double stranded DNA breaks induced by aberrant topoisomerase II activity (PEDERSEN-BJERGAARD and ROWLEY 1994). An insufficient number of t-AML breakpoints, however, have been sequenced to draw any conclusions about such potential mechanisms at present.

3.4 Proposed Mechanisms of Illegitimate Recombination Involving *HRX*

On the basis that *HRX* breakpoints in some de novo leukemias have been found to lie within or near to putative topo II and V-D-J recombinase recognition sites or *Alu* repeats or between SARs (GU et al. 1992b, 1994; DOMER et al. 1993, 1995; NEGRINI et al. 1993; SCHICHMAN et al. 1994a,b; BROEKER et al. 1996), it has been suggested that the *HRX* breakpoint region might be susceptible to illegitimate somatic recombination as a result of aberrant topoisomerase II or recombinase enzyme activity or because of some peculiarity of DNA structure in that region. It is noteworthy, however, that such recognition sites are absent in the majority of de novo HRX-associated leukemias and that such sequences have in general proved to be poor matches to the topo II consensus or imperfectly spaced for proper recognition by the recombinase enzyme. Likewise, *Alu* – mediated homologous recombination is an unlikely cause of reciprocal translocations involving *HRX* because the disrupted *Alu* elements do not reconstruct chimeric full length *Alu* sequences (Gu et al. 1994), although such a mechanism may underlie partial duplications of *HRX* (Schichman et al. 1994b).

4 HRX Fusion Proteins

4.1 HRX Fusion Proteins

Several HRX translocation partners have been recently cloned as cDNAs and characterized (Gu et al. 1992b; Tkachuk et al. 1992; Corral et al. 1993; Iida et al. 1993b; Morrisey et al. 1993; Nakamura et al. 1993; Prasad et al. 1993, 1994; McCabe et al. 1994; Parry et al. 1994; Thirman et al. 1994; Chaplin et al. 1995a; Mitani et al. 1995; Tse et al. 1995; Marschalek et al. 1995). They comprise a diverse group that is summarized in Table 1. Some are structurally related (e.g., ENL and AF-9, AF-10 and AF-17, and AF-6 and AF-1p) (Fig. 3), and all have in common the fact that they encode novel, probably nuclear, proteins that are normally expressed in hematopoietic cells. Furthermore in all cases the HRX fusion transcripts are predicted to form in-frame chimeric proteins.

ENL and AF-9 are 82% identical at their NH_2- and COOH-terminals (Nakamura et al. 1993; Iida et al. 1993b; Rubnitz et al. 1994b) and share this same structural homology with a yeast protein, Ancl, a nuclear protein with diverse roles (see below) (Welch and Drubin 1994; Rubnitz et al. 1994b). ENL and FEL both possess potential nuclear targeting motifs, several potential tyrosine phosphorylation sites and potential ATP/GTP or GTP-binding sites (Tkachuk et al. 1992; Yamamoto et al. 1993b; Morrissey et al. 1993; Domer et al. 1993; Nakamura et al. 1993), and are capable of activating transcription in transient cellular assays (Rubnitz et al. 1994b; Morrissey and Cleary, unpublished). AF-10 (Chaplin et al. 1995a) and AF-17 (Prasad et al. 1994) both contain NH_2-terminal triplets of imperfect PHD fingers (93% identical), similar to those of HRX (Aasland et al. 1995), and COOH-terminal leucine zipper (ZIP) (77% identical) motifs that lack the sequence-specific DNA-binding basic region of bZIP proteins suggestive of a role in protein dimerization. Both motifs are conserved in sequence and position with those of a human bromodomain-containing protein, BR140, that is homologous to the TAF250 subunit of TFIID (Thompson et al. 1994). AF-1p is closely

Table 1. Features of HRX translocation partners

Name	Locus	Leukemia	Protein amino acids	Location	Motifs	Homology	Expressed	References
AF-1p.	1p32	AML-M0	896	Cytoplasmic(?)	α HCC	Murine eps15 Yeast poly(A) ribonuclease. Myosin	–	Bernard et al. 1994
AF1q	1q21	AML-M4	90	Nuclear	–	–	B,T thymus Myeloid	Tse et al. 1995
FEL (AF-4)	4q21	B ALL	1210	Nuclear	Nuclear targeting motif GTP-binding motif	–	Ubiquitous T and B cells	Gu et al. 1992b Morrissey et al. 1993 Chen et al. 1993a
AF-6	6q27	AML-M4	1612	Cytoplasmic(?)	GLGF repeat α HCC	PSD95 ZO-1 Drosophila dlg Yeast myosin-1 *D.discoideum* myosin heavy chain	Ubiquitous	Prasad et al. 1993.
AF-9 (LTG9) (MLLT3)	9p22	B-ALL AML	568	Nuclear	Nuclear targeting motif	ENL Anc 1 RNA polymerase II	Lymphoid organs Megakaryocytic Erythroid	Nakamura et al. 1993 Iida et al. 1993b Rubnitz et al. 1994b
AF-10	10p12	AML-M5	1027	Nuclear	PHD fingers Leucine zipper	AF-17 Br140 *C.elegans* CEZF	Ubiquitous Testes	Chaplin et al. 1995a
AF17	17q21	AML	1093	Nuclear	PHD fingers Leucine zipper	AF-10 Br140	Ubiquitous	Prasad et al. 1994

ELL (MEN)	19p13.1	AML	621	Nuclear	Highly basic, lysine-rich	Poly (ADP-ribose) polymerase Arginine-rich nuclear protein *Engrailed* gene Knob protein of *P. falciparum* Bovine leukemia virus receptor	Ubiquitous	Thirman et al. 1994 Mitani et al. 1995
ENL (LTG19) (MLLT1)	19p13.3	B-ALL	559	Nuclear	Nuclear targeting motif ATP/GTP-binding site Phosophorylation sites	AF-9 Anc1 RNA polymerase II	Ubiquitous	Tkachuk et al. 1992 Nakamura et al. 1993 Iida et al. 1993b Yamamoto et al. 1993b Rubnitz et al. 1994b
AF-X1	Xq13	T-ALL	?	Nuclear	Forkhead domain	ALV(FKHR) Avian retroviral oncogene, *qin*. Interleukin binding factor Human hepatocyte nuclear factor-3a	Ubiquitous	Corral et al. 1993, Parry et al. 1994 McCabe et al. 1994

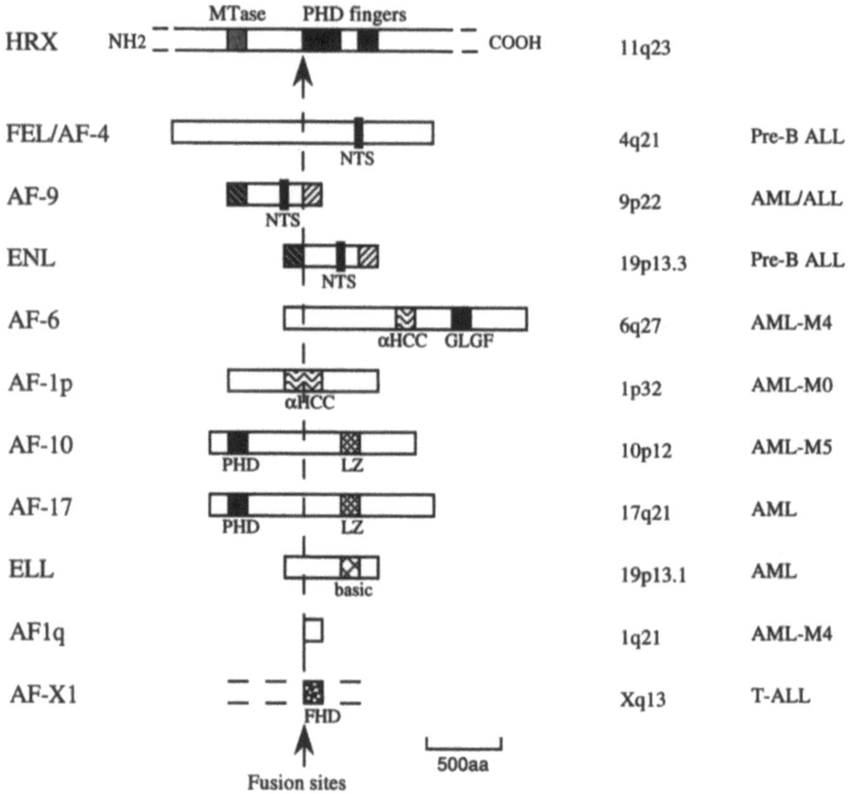

Fig. 3. The structure of the ten cloned unrearranged HRX-fusion partners, their chromosomal location and disease association. The leukemogenic chimeric proteins are thought to be encoded on the derivative (11) chromosome in 11q23 reciprocal translocations and comprise the NH_2-terminal portion of HRX fused to the COOH-terminal portions of the various translocation partners. The arrows and vertical dashed line represent the fusion point which uncouples the methyltransferase (MTase) and PHD finger domains of HRX. *NTS* nuclear targeting sequence; *αHCC*, α-helical coiled-coil; *GLGF*, GLGF repeat; *LZ*, lucine zipper; basic, basic amino acid region; *FHD*, forkhead domain. Note: AF-X1 has only been partially sequenced and is represented by *horizontal dashed lines*

related to poly(A) nuclease (PAN) (47% identity) (Bernard et al. 1994), which is involved in posttranscriptional modification of mRNA by shortening poly(A) tails (Sachs and Deardorff 1992), and murine *eps15* (epidermal growth factor receptor pathway substrate 15) (88% similarity) (Bernard et al. 1994), which is a cytoplasmic phosphoprotein that is a target for the tyrosine kinase activity of the EGF receptor and, notably, has known transforming ability (Fazioli et al. 1993). AF-6 and AF-1p both contain heptad repeats (Prasad et al. 1993; Bernard et al. 1994) typical of the α-helical coiled-coil structure of various cytoskeletal (e.g., myosin heavy chains) and nuclear transcription factors (e.g., PML). ELL contains a highly basic, lysine-rich, domain that is homologous to similar regions of other DNA-binding proteins (Thirman et al. 1994; Mitani et al. 1995). AFX1 shows strong homology within its forkhead box to members of the forkhead transcription factor family, particularly ALV (FKHR) (Corral et al. 1993; Parry et al. 1994). Curiously, AFX1 is bisected at a comparable site as is FKHR in the t(2;13)(q35;q14) translocation in alveolar rhabdomyosarcoma that forms a PAX3-FKHR chimera (Galili et al. 1993).

4.2 Pathogenic Role of HRX Fusion Proteins in Acute Leukemia

Several lines of evidence indicate that it is the NH_2-HRX-partner-COOH product encoded by the derivative (11) chromosome that is etiologically important in leukemias. First, the der (11) product is conserved in all complex 11q23 translocations (Rowley 1992); second, the der (11) product is always expressed even when the reciprocal product is not (Morrissey et al. 1993; Chen et al. 1993a; Borkhardt et al. 1994; Janssen et al. 1994; Downing et al. 1994); and third, 3', but not 5', HRX sequences are frequently deleted from the breakpoint region in *HRX* translocations (Thirman et al. 1993a,b). Given these observations, 11q23-induced chimeras are predicted to retain the NH_2-terminal portion of HRX containing the minor groove DNA-binding motifs, transcriptional repression domain, and the snRNP, MTase and histone H1 homology regions. Leukemogenic HRX chimeric proteins would not retain the more highly conserved PHD and SET domains, present in Drosophila *trithorax* and *Polycomb*-group proteins, and would lose the transcriptional activation domain. Furthermore, der(11) chimeric HRX fusion proteins would be expressed under control of the, as yet, undefined regulatory regions that govern expression of the unrearranged *HRX* gene, which is active in hematopoietic cells.

A major unresolved issue concerns whether the various HRX-chimeric proteins act through a gain-of-function or loss-of-function mechanism, and there are compelling arguments for both alternatives. The case for gain-of-function arises from observations that all known HRX chimeric proteins display in-frame fusions, despite the occurrence of deletions (Thirman et al. 1993a,b), inverted repeats (Domer et al. 1993), and addition of N- region nucleotides (Negrini et al. 1993) at breakpoints. Furthermore, although quite heterogeneous, most of the presently characterized partner proteins possess features suggestive of a transcriptional role and some are closely related or share common structural features. Thus, there appear to be strong selective pressures for specific functional contributions by the various partner proteins with some contributing their potential trans-activation domains (e.g., FEL, ENL, AF-9, and AF-X1), while others donate their dimerization motifs (e.g. AF-10, and AF-17), or protein-protein interaction domains (e.g. AF-1p and AF-6). However, given the consistent inclusion of the NH_2-terminal portion of HRX in 11q23 chimeras, its functional contributions appear to be identical in all forms of leukemia and most likely relates to its nonclassical DNA recognition motifs. These may serve to inappropriately localize the various fusion partners to genomic sites normally occupied by wild-type HRX. Opponents of this model argue, convincingly, that given the variability of the multiple HRX fusion partners that do not appear to contribute any consistent structural features, it is difficult to imagine how so many unrelated partners could contribute a common transforming function to the chimera.

The loss-of-function hypothesis contends that the sole consistent event in all *HRX* translocations is disruption of HRX which alone may be a sufficient oncogenic mutation. Support for this view comes from the leukemias that possess partial tandem internal duplications of HRX in which a contribution is clearly not required from a heterologous fusion partner. Additional support for the loss-of-function hypothesis comes from the observation that HRX+/– mice display hematologic abnormalities indicating that expression of both *HRX* alleles are essential for normal hematopoiesis and that dosage is important. Thus, a contributing event in HRX-associated leukemogenesis may be functional loss of at least one allele. However, a simple loss-of-function mechanism does not adequately account for the consistent in-frame fusions invariably produced by 11q23 translocations. An intriguing possibility is that *HRX* translocations produce a "double hit," by simultaneously inducing haplo-insufficiency for wild-type HRX and a gain-of-function mutation for the translocated HRX protein, but how this may contribute to leukemogenesis is unknown.

5 Clues to the Function of Wild-Type HRX

A growing body of structural and genetic evidence supports an association between trx family proteins and chromatin structure that may shed some light on the normal function of HRX and give some insight into the leukemogenic consequences of its disruption. Firstly, the PHD fingers and SET domains of HRX and trx are also present in other proteins implicated in the maintenance of homeotic gene expression patterns, e.g., Pcl (Lonie et al. 1994; Aasland et al. 1995), and the regulation of chromatin structure, e.g., Su(var)3–9 and E(z) (Tschiersch et al. 1994; Rastelli et al. 1993). Notably, mutations of the *E*(*z*) gene resulted in disruption of the *Pc*-G complex and cause decondensation of chromatin (Rastelli et al. 1993), and *Su*(*var*)3–9 is a suppressor of PEV in *Drosophila* (Tschiersch et al. 1994).

Secondly, the NH_2-terminal of HRX displays similarity to functional domains of some chromatin-associated proteins such as histones and HMG-I(Y) proteins. The AT hook domains of HMG proteins are required to establish an open chromatin structure by competing with histone H1 and/or nucleosomes for binding to AT-rich regulatory sequences (Reeves 1992), raising the possibility of a similar function for HRX. This may also apply to the SPKK motifs and the lysine-rich regions that are similar to those present in the minor groove-binding tails of some histones. If HRX does indeed compete for such sites, then a likely competitor would be histone H1 that has a general repressive effect on transcription (Wolfe 1991, 1994; Croston et al. 1991). The AT hook region of HRX may also act at a higher order of chromatin structure since this region also binds to SARs that organize interphase and mitotic chromatin into a series of loop domains (Gasser et al. 1989).

Thirdly, the snRNP and MTase-like regions of HRX, which are similar to that part of mammalian DNA MTase that is able to distinguish hemimethylated from methylated CpG sites, might be able to read the methylation status of the promoter sequences within the target genes of HRX. Methylation patterns, established during early embryogenesis and gametogenesis, undergo characteristic changes during differentiation presumably in association with activation of many tissue-specific genes (CEDAR 1988). Such patterns govern gene activity and are determined by epigenetic factors (CEDAR 1988; LI et al. 1992). HRX may, therefore, provide a link between methylation patterns and chromatin structure (BESTOR and VERDINE 1994), both of which are maintained in a clonally inherited, tissue-specific fashion throughout development (LI et al. 1992; PARO et al. 1993).

Genetic studies of other *Trx*-G genes in *Drosophila* and their homologues in yeast have provided some functional evidence of how HRX might work. A member of the *Trx*-G, trithorax-like (*Trl*), encodes the *Drosophila* GAGA factor which is required for the maintenance of homeotic gene expression and is an enhancer of PEV (FARKAS et al. 1994). GAGA binding sites are found in the promoter regions of several *Drosophila* genes and GAGA factor has been shown to bind in a site-specific manner to the promoter of *Ubx* and stimulates its transcription in vitro (BIGGIN and TIJAN 1988). GAGA factor acts as an anti-repressor by counteracting the inhibitory activity of histone H1 by preventing its nonspecific binding in critical regions of target promoters (CROSTON et al. 1991) and by generating, and presumably maintaining, nucleosome-free regions of DNA at promoters (TSUKIYAMA et al. 1994). Another *Trx*-G protein, brahma (brm), is homologous to the yeast protein, SW12 (TAMKUN et al. 1992; KENNISON 1993), and its human counterpart, BRG1 (KHAVARI et al. 1993). These proteins are part of highly conserved, multicomponent complexes (called SWI/SNF) that may enhance transcription by directly modifying chromatin structure (CAIRNS et al. 1994; CÔTÉ et al. 1994; KWON et al. 1994; IMBALZANO et al. 1994). Complexes composed of members of the SWI/SNF family facilitate transcription by antagonizing the repressive effect of chromatin by interacting with histones and HMG-like proteins (WINSTON and CARLSON 1992) and by directly inducing alterations in nucleosomal structure (CÔTÉ et al. 1994; KWON et al. 1994; IMBALZANO et al. 1994). Genetic experiments in *Drosophila* suggest that *brahma* assists homeotic activators such as *trx* in relieving the repressive effect of chromatin, perhaps through a *Drosophila* counterpart of the SWI/SNF complex (TAMKUN et al. 1992). Like HMG proteins and GAGA factor, HRX may act by preventing histone H1 and other chromatin-associated proteins from incorporating into and condensing nucleosomes of the regulatory regions of otherwise repressed genes, perhaps, in part, by interacting with a human SWI/SNF complex.

6 Clues to the Leukemogenic Action of HRX Chimeras

Somewhat unexpectedly, some of the HRX fusion partners may also be components of, or interact with, chromatin-modifying protein complexes. The yeast protein Ancl, which is homologous to ENL and AF-9, has been shown to be present in the yeast SWI/SNF complex (Cairns B, personal communication) and is a component of two basal transcription complexes, TFIID (TAF30) and TFIIF (Tfg3) (Henry et al. 1994). It is not yet known whether ENL and AF-9 are present in, or interact with, the basal transcriptional complexes or the human SWI/SNF complex that contains BRG-1. In addition, BR140, a relative of AF-10 and AF-17, is homologous to the TAF250 subunit of TFIID and belongs to the bromodomain-containing family that also includes SWI2, brm, and BRG-1 (Thompson et al. 1994). These observations suggest that ENL, AF-9, AF-10, AF-17 and potentially other proteins fused to HRX in leukemias may be members of, or interact with, the human SWI/SNF multimeric complex and/or the basal transcriptional apparatus. If this is the case, a gain-of-function could result from replacement of the COOH-terminal portion of HRX with those regions of the partner proteins that are essential for interaction with the SWI/SNF complex and/or basal transcriptional apparatus. Consequently, the normal regulated interactions of HRX with the SWI/SNF or basal transcriptional complexes would be short-circuited allowing their constitutive localization to genomic sites by the NH_2-terminal portion of HRX. This may contribute to inappropriate expression or suppression of target genes, depending upon which factors were recruited into the chimeric HRX complexes. Alternatively, the partner proteins may, as a result of their fusion to HRX, be sequestered from the SWI/SNF and/or basal transcription complexes, leading to their functional compromise in support of a loss-of-function model.

In either model, HRX dosage is reduced (which is sufficient to induce homeotic mutations) but not completely abolished. However, HRX chimeras could theoretically have dominant-negative properties and heterodimerize with wild-type HRX (or wild-type fusion partners) resulting in additional functional impairment. Whatever the mechanism, early hematopoietic cells bearing HRX mutations would fail to maintain, or appropriately alter, the correct chromatin pattern of *Hox* gene (or other) promoter regions during subsequent differentiation, with consequent altered *Hox* gene expression and disordered cell fate. Alterations to the fate of primitive hematopoietic cells are presumably coupled to continued proliferation, which likely renders them susceptible to further leukemogenic events. Additional events may include mutations in p53 or ras, or dysregulated bcl-2 function, which have been documented to accompany *HRX* rearrangements in some cases (Lanza et al. 1995; Pocock et al. 1995; Naoe et al. 1993).

7 Conclusion

Recent work on transcriptional regulatory pathways in yeast, the genetic basis for the maintenance of the active or repressed state of *Hox* gene expression in *Drosophila*, and the biochemical basis underlying the structure of chromatin have all been converging to provide new insights into the complex world of higher order gene regulation. Many of the structural features of HRX and its relationship to trithorax seem to touch upon aspects of this still incomplete picture. The unifying theme that emerges from considering these, as yet unproved, associations is that HRX may be a new type of target-specific epigenetic regulatory factor that is important in the recognition and maintenance of higher order chromatin structure. By studying the effects of disruption of HRX in leukemias, we predict that an entirely new mechanism of tumorigenesis will be revealed.

Acknowledgements. The authors would like to thank Xiangmin Cui and Robert Slany for valuable discussions in the preparation of this manuscript and Brad Cairns and Nancy Zeleznik-Le for sharing their unpublished data. PMW is a recipient of a Lucille P. Markey Australian Visitors Fellowship. MLC is a scholar of the Leukemia Society of America.

References

Aasland R, Gibson TJ, Stewart AF (1995) The PHD finger: implications for chromatin-mediated transcriptional regulation. Trends Biochem Sci 20: 56–59

Alkema MJ, van der Lugt NMT, Bobeldijk RC, Berns A, van Lohuizen M (1995) Transformation of axial skeleton due to overexpression of *bmi-1* in transgenic mice. Nature 374: 724–727

Bestor TH (1992) Activation of mammalian DNA methyltransferase by cleavage of a ZN binding regulatory domains. EMBO J 11: 2611–2617

Bestor T, Laudano A, Mattaliano R, Ingram V (1988) Cloning and sequencing of a cDNA encoding DNA methyltransferase of mouse cells. The carboxy-terminal domain of the mammalian enzymes is related to bacterial restriction methyltransferases. J Mol Biol 203: 971–983

Bestor TH, Verdine GL (1994) DNA methyltransferases. Curr Biol 6: 380–389

Bernard OA, Mauchauffe M, Mecucci C, Van den Berghe H, Berger R (1994) A novel gene, AF-1p, fused to HRX in t(1;11)(p32;q23), is not related to AF-4, AF-9 nor ENL. Oncogene 9: 1039–1045

Biggin M, Tijan R (1988) Transcription factors that activate the *Ultrabithorax* promoter in developmentally staged extracts. Cell 53: 699–711

Borkhardt A, Repp R, Haupt E, Brettreich S, Buchen U, Gossen R, Lampert F (1994) Molecular analysis of MLL-1/AF4 recombination in infant acute lymphoblastic leukemia. Leukemia 8: 549–553

Bower M, Parry P, Carter M, Lillington DM, Amess J, Lister TA, Evans G, Young BD (1994a) Prevalence and clinical correlations of *MLL* gene rearrangements in AML-M4/5. Blood 84: 3776–3780

Bower M, Parry P, Gibbons B, Amess J, Lister TA, Young BD, Evans GA (1994b) Human trithorax gene rearrangements in therapy-related acute leukaemia after etoposide treatment. Leukemia 8: 226–229

Breen TR, Harte PJ (1991) Molecular characterization of the *trithorax* gene, a positive regulator of homeotic gene expression in *Drosophila*. Mech Dev 35: 113–127

Breen TR, Harte PJ (1993) *trithorax* regulates multiple homeotic genes in the bithorax and Antennapedia complexes and exerts different tissue-specific, parasegment-specific and promoter-specific effects on each. Development 117: 119–134

Broeker PL, Gill Super H, Thirman MJ, Pomykala H, Yuka Y, Tanabe S, Zeleznik-Le N, Rowley J (1994) Distribution of 11q23 breakpoints within the *MLL* breakpoint cluster region in de novo acute leukemia and in treatment related acute myeloid leukemia: correlation with scaffold attachment regions and topoisomerase II consensus binding sites. Blood 87: 1912–1922

Broeker P, Harden A, Rowley J, Zeleznik-Le N (1995) The mixed lineage leukemia (MLL) protein involved in 11q23 transloactions contains a domain that binds cruciform DNA and scaffold attachment region (SAR) DNA. Curr Topic Micro Immunol (in press)

Brunk BP, Martin EC, Adler PN (1991) Drosophila genes *Posterior sex combs* and *Suppressor two of zeste* encode proteins with homology to the murine *bmi-1* oncogene. Nature 353: 351–353

Cairns BR, Kim YJ, Sayre MH, Laurent BC, Kornberg RD (1994) A multisubunit complex containing the *SWI1/ADR6, SWI2/SNF2, SWI3, SNF5,* and *SNF6* gene products isolated from yeast. Proc Natl Acad Sci USA 91: 1950–1954

Castelli-Gair JE, García-Bellido A (1990) Interactions of *Polycomb* and *trithorax* with *cis* regulatory regions of *Ultrabithorax* during the development of *Drosophila melanogaster*. EMBO J 9: 4267–4275

Cedar H (1988) DNA methylation and gene activity. Cell 53: 3–4

Chan C-S, Rastelli L, Pirrotta V (1994) A *Polycomb* response element in the *Ubx* gene that determines an epigenetically inherited state of repression. EMBO J 13: 2553–2564

Chaplin TC, Ayton P, Bernard OA, Saha V, Della Valle V, Hillion J, Gregorini A, Lillington D, Berger R, Young BD (1995a) A novel class of zinc finger/leucine zipper genes identified from the molecular cloning of the t(10;11) translocation in acute leukemia. Blood 85: 1435–1441

Chaplin TC, Bernard OA, Beverloo HB, Saha V, Hagemeijer A, Berger R, Young BD (1995b) The t(10;11) translocation in acute myeloid leukemia (M5) consistently fuses the leucine zipper motif of AF10 onto the HRX gene. Blood 86: 2073–2076

Chen CS, Medberry PS, Arthur DC, Kersey JH (1991) Breakpoint clustering in t(4;11) (q21;q23) acute leukemia. Blood 78: 2498–2504

Chen CS, Hilden JM, Frestedt J, Domer PH, Moore R, Korsmeyer SJ, Kersey JH (1993a) The chromosome 4q21 gene (AF-4/FEL) is widely expressed in normal tissues and shows breakpoint diversity in t(4;11)(q21;q23) acute leukemia. Blood 82: 1080–1085

Chen CS, Sorensen PH, Domer PH, Reaman GH, Korsmeyer SJ, Heerema NA, Hammond GD, Kersey JH (1993b) Molecular rearrangements on chromosome 11q23 predominate in infant acute lymphoblastic leukemia and are associated with specific biologic variables and poor outcome. Blood 81: 2386–2393

Churchill M, Susuki M (1989) "SPKK" motifs prefer to bind DNA at AT-rich sites. EMBO J 8: 4189–4195

Cimino G, Moir DT, Canaani O, Williams K, Crist WM, Katzav S, Cannizzaro L, Lange B, Nowell PC, Croce CM, Canaani E (1991) Cloning of ALL-1, the locus involved in leukemias with the t(4;11)(q21;q23), t(9;11)(p22;q23), and t(11;19)(q23;p13) chromosome translocations. Cancer Res 51: 6712–6714

Cimino G, Nakamura T, Gu Y, Canaani O, Prasad P, Crist WM, Carroll AJ, Baer M, Bloomfield CD, Nowell PC, Croce CM, Canaani E (1992) An altered 11-kilobase transcript in leukemic cell lines with the t(4;11)(q21;q23) chromosome translocation. Cancer Res 52: 3811–3813

Cimino G, Lo Coco F, Biondi A, Elia L, Luciano A, Croce CM, Masera G, Mandelli F, Canaani E (1993) *ALL-1* gene at chromosome 11q23 is consistently altered in acute leukemia of early infancy. Blood 82: 544–546

Cimino G, Rapanotti MC, Elia L, Biondi A, Fizzotti M, Testi AM, Tosti S, Croce CM, Canaani E, Mandelli F, Lo Coco F (1995a) *ALL-1* gene rearrangements in acute myeloid leukemia: association with M4-M5 French-American-British classification subtypes and young age. Cancer Res 55: 1625–1628

Ciminio G, Rapanotti MC, Rivolta A, Lo Coco F, D'Arcangelo E, Rondelli R, Basso G, Barisone E, Rosanda C, Santostasi T, Canaani E, Masera G, Mandelli F, Biondi A (1995b) Prognostic relevance of ALL-1 gene rearrangement in infant acute leukemias. Leukemia 9: 391–395

Corral J, Forster A, Thompson S, Lampert F, Kaneko Y, Slater R, Kroes WG, van der Schoot CE, Ludwig WD, Karpas A, Pocock C, Cotter F, Rabbitts TH (1993) Acute leukemias of different lineages have similar MLL gene fusions encoding related chimeric proteins resulting from chromosomal translocation. Proc Natl Acad Sci USA 90: 8538–8542

Côté J, Quinn J, Workman JL, Peterson CL (1994) Stimulation of GAL4 derivative binding to nucleosomal DNA by the yeast SWI/SNF complex. Science 265: 53–60

Croston GE, Kerrigan LA, Lira LM, Marshak DR, Kadonaga JT (1991) Sequence-specific antirepression of histone H1-mediated inhibition of basal RNA polymerase II transcription. Science 251: 643–649

Djabali M, Selleri L, Parry P, Bower M, Young BD, Evans GA (1992) A trithorax-like gene is interrupted by chromosome 11q23 translocations in acute leukaemias. Nat Gen 2: 113–118

Domer PH, Fakharzadeh SS, Chen CS, Jockel J, Johansen L, Silverman GA, Kersey JH, Korsmeyer SJ (1993) Acute mixed-lineage leukemia t(4;11)(q21;q23) generates an *MLL-AF4* fusion product. Proc Natl Acad Sci USA 90: 7884–7888

Domer PH, Head DR, Renganathan N, Raimondi SC, Yang E, Atlas M (1995) Molecular analysis of 13 cases of *MLL*/11q23 secondly acute leukemia and identification of topoisomerase II consensus-binding sequences near chromosome breakpoint of a secondary leukemia with the t(4;11). Leukemia 9: 1305–1312

Downing JR, Head DR, Raimondi SC, Carroll AJ, Curcio-Brint AM, Motroni TA, Hulshof MG, Pullen DJ, Domer PH (1994) The der(11)-encoded MLL/AF-4 fusion transcript is consistently detected in t(4;11)(q21;q23)-containing acute lymphoblastic leukemia. Blood 83: 330–335

Farkas G, Gausz J, Galloni M, Reuter G, Gyurkovics H, Karch F (1994) The *Trithorax-like* gene encodes the *Drosophila* GAGA factor. Nature 371: 806–808

Fazioli F, Minichiello L, Matoskova B, Wong WT, Di Fiore P (1993) eps15, a novel tyrosine kinase substrate, exhibits transforming activity. Mol Cell Biol 13: 5814–5828

Felix CA, Winick NJ, Negrini M, Bowman WP, Croce CM, Lange BJ (1993) Common region of *ALL-1* gene disrupted in epipodophyllotoxin-related secondary acute myeloid leukemia. Cancer Res 53: 2954–2956

Felix CA, Hosler MR, Winick NJ, Masterson M, Wilson AE, Lange BJ (1995) *ALL-1* gene rearrangements in DNA topoisomerase II inhibitor-related leukemia in children. Blood 85: 3250–3256

Ford AM, Ridge SA, Cabrera ME, Mahmoud H, Steel CM, Chan LC, Greaves M (1993) In utero rearrangements in the trithorax-related oncogene in infant leukaemias. Nature 363: 358–360

Galili N, Davis RJ, Fredericks WJ, Mukhopadhyay S, Rauscher FJ III, Emanuel BS, Rovera G, Barr FG (1993) Fusion of fork head domain gene to PAX3 in solid tumor alveolar rhabdomyosarcoma. Nat Gen 5: 230–235

Gasser SM, Amati BB, Cardenas ME, Hoffman JF-X (1989) Studies on scaffold-attachment sites and their relation to genome function. In: Jeon KW, Friedlander M (eds) Int Rev Cytol Acad Press, New York

Gill Super HJ, McCabe NR, Thirman MJ, Larson RA, Le Beau MM, Pedersen-Bjergaard J, Philip P, Diaz MO, Rowley JD (1993) Rearrangements of the *MLL* gene in theraphy-related acute myeloid leukemia in patients previously treated with agents targeting DNA-topoisomerase II. Blood 82: 3705–3711

Gill Super HJ, Rothberg PG, Kobayashi H, Freeman AI, Diaz MO, Rowley JD (1994) Clonal, non-constitutional rearrangements of the *MLL* gene in infant twins with acute lymphoblastic leukemia: in utero chromosome rearrangement of 11q23. Blood 83: 641–644

Gindhart JG Jr, Kaufman TC (1995) Identification of *Polycomb* and *trithorax* group responsive elements in the regulatory region of the Drosophila homeotic gene *Sex combs reduced*. Genetics 139: 797–814

Gu Y, Cimino G, Alder H, Nakamura T, Prasad R, Canaani O, Moir DT, Jones C, Nowell PC, Croce CM, Canaani E (1992a) The (4;11)(q21;q23) chromosome translocations in acute leukemias involve the VDJ recombinase. Proc Natl Acad Sci USA 89: 10464–10468

Gu Y, Nakamura T, Alder H, Prasad R, Canaani O, Cimino G, Croce CM, Canaani E (1992b) The t(4;11) chromosome translocation of human acute leukemias fuses the *ALL-1* gene, related to Drosophila *trithorax*, to the *AF-4* gene. Cell 71: 701–708

Gu Y, Alder H, Nakamura T, Schichman SA, Prasad R, Canaani O, Saito H, Croce CM, Canaani E (1994) Sequence analysis of the breakpoint cluster region in the *ALL-1* gene involved in acute leukemia. Cancer Res 54: 2326–2330

Height SE, Dainton MG, Kearney L, Swansbury GJ, Matutes E, Dyer MJ, Treleaven JG, Powles RL, Catovsky D (1994) Acute myelomonocytic leukemia with t(10;11)(p13;q23): heterogeneity of breakpoints at 11q23 and association with recombinase activation. Genes Chromosome Cancer 11: 136–139

Henry NL, Campbell AM, Feaver WJ, Poon D, Weil PA, Kornberg RD (1994) TFIIF-TAF-RNA polymerase II connection. Genes Dev 8: 2868–2878

Hernandez JM, Mecucci C, Beverloo HB, Selleri L, Wlodarska I, Stul M, Michaux L, Verhoef G, Van Orshoven A, Cassiman JJ, Evans GA, Hagemeijer A, Van den Berghe H (1995) Translocation (11;15)(q23;q14) in three patients with acute non-lymphoblastic leukemia (ANLL); clinical, cytogenetic and molecular studies. Leukemia 9: 1162–1166

Hilden JM, Chen CS, Moore R, Frestedt J, Kersey JH (1993) Heterogeneity in MLL/AF-4 fusion messenger RNA detected by the polymerase chain reaction in t(4;11) acute leukemia. Cancer Res 53: 3853–3856

Hunger SP, Tkachuk DC, Amylon MD, Link MP, Carroll AJ, Welborn JL, Willman CL, Cleary ML (1993) *HRX* involvement in de novo and secondary leukemias with diverse chromosome 11q23 abnormalities. Blood 81: 3197–3203

Iida S, Seto M, Yamamoto K, Komatsu H, Akao Y, Nakazawa S, Ariyoshi Y, Takahashi T, Ueda R (1993a) Molecular cloning of 19p13 breakpoint region in infantile leukemia with t(11;19)(q23;p13) translocation. Jpn J Cancer Res 84: 532–537

Iida S, Seto M, Yamamoto K, Komatsu H, Tojo A, Asano S, Kamada N, Ariyoshi Y, Takahashi T, Ueda R (1993b) *MLLT3* gene on 9p22 involved in t(9;11) leukemia encodes a serine/proline rich protein homologous to MLLT1 on 19p13. Oncogene 8: 3085–3092

Imbalzano AN, Kwon H, Green MR, Kingston RE (1994) Facilitated binding of TATA-binding protein to nucleosomal DNA. Nature 370: 481–485

Jani Sait SN, Raimondi SC, Look AT, Gill H, Thirman M, Diaz MO, Shows TB (1993) A t(11;12) 11q23 leukemic breakpoint that disrupts the *MLL* gene. Genes Chromosom Cancer 7: 28–31

Janssen JWG, Ludwig WD, Borkhardt A, Spadinger U, Rieder H, Fontasch C, Hossfeld DK, Harbott J, Schulz AS, Repp R, Sykora KW, Hoelzer D, Bartram CR (1994) Pre-pre-B acute lymphoblastic leukemia: high frequency of alternatively spliced ALL1-AF4 transcripts and absence of minimal residual disease during complete remission. Blood 84: 3835–3842

Jones RS, Gelbart WM (1993) The *Drosophila* polycomb-group gene *Enhancer of zeste* contains a region with sequence similarity to *trithorax*. Mol Cell Biol 13: 6357–6366

Kearney L, Bower M, Gibbons B, Das S, Chaplin T, Nacheva E, Chessells JM, Reeves B, Riley JH, Lister TA, Young BD (1992) Chromosome 11q23 translocations in both infant and adult acute leukemias are detected by in situ hybridization with a yeast artificial chromosome. Blood 80: 1659–1665

Kennison JA (1993) Transcriptional activation of *Drosophila* homeotic genes from distant regulatory elements. Trends Genet 9: 75–79

Khavari PA, Peterson CL, Tamkun JW, Mendel DB, Crabtree GR (1993) BRG1 contains a conserved domain of the *SWI2/SNF2* family necessary for normal mitotic growth and transcription. Nature 366: 170–174

Kobayashi H, Espinosa R, Thirman MJ, Gill HJ, Fernald AA, Diaz MO, Le Beau MM, Rowley JD (1993) Heterogeneity of breakpoints of 11q23 rearrangements in hematologic malignancies identified with fluorescence in situ hybridization. Blood 82: 547–551

Krumlauf R (1994) *Hox* genes in vertebrate development. Cell 78: 191–201

Kuzin B, Tillib S, Sedkov Y, Mizrokhi L, Mazo A (1994) The *Drosophila trithorax* gene encodes a chromosomal protein and directly regulates the region-specific homeotic gene *fork head*. Genes Dev 8: 2478–2490

Kwon H, Imbalzano AN, Khavari PA, Kingston RE, Green MR (1994) Nucleosome disruption and enhancement of activator binding by a human SWI/SNF complex. Nature 370: 477–481

Lanza C, Gaidano G, Cimino G, Lo Coco F, Basso G, Sainati L, Pastore C, Nomdedeu J, Volpe G, Parvis G, Barisone E, Mazza U, Madon E, Saglio G (1995) *p53* gene inactivation in acute lymphoblastic leukemia of B cell lineage associates with chromosomal breakpoints at 11q23 and 8q24. Leukemia 9: 955–959

Leblanc T, Hillion J, Derre J, Le Coniat M, Baruchel A, Daniel MT, Berger R (1994) Translocation t(11;11)(q13;q23) and HRX gene rearrangement associated with therapy-related leukemia in a child previously treated with VP16. Leukemia 8: 1646–1648

Leonhardt H, Page AW, Weier H-U, Bestor TH (1992) A targeting sequence directs DNA methyltransferase to sites of DNA replication in mammalian nuclei. Cell 71: 865–873

Li E, Bestor TH, Jaenisch R (1992) Targeted mutation of the DNA methyltransferase gene results in embryonic lethality. Cell 69: 915–926

Lonie A, D'Andrea R, Paro R, Saint R (1994) Molecular characterization of the *Polycomblike* gene of *Drosophila melanogaster*, a *trans*-acting negative regulator of homeotic gene expression. Development 120: 2629–2636

Ma Q, Alder H, Nelson KK, Chatterjee D, Gu Y, Nakamura T, Canaani E, Croce CM, Siracusa LD, Buchberg AM (1993) Analysis of the murine *All-1* gene reveals conserved domains with human *ALL-1* and identifies a motif shared with DNA methyltransferases. Proc Natl Acad Sci USA 90: 6350–6354

Marschalek R, Greil J, Lochner K, Nilson I, Siegler G, Zweckbronner I, Beck JD, Fey GH (1995) Molecular analysis of the chromosomal breakpoint and fusion transcripts in the acute lymphoblastic SEM cell line with chromosomal translocation t(4;11). Br J Haematol 90: 308–320

Martinex-Climent JA, Thirman MJ, Espinosa R III, Le Beau MM, Rowley JD (1995) Detection of 11q23/*MLL* rearrangements in infant leukemias with fluorescence *in situ* hybridization and molecular analysis. Leukemia 9: 1299–1304

Mazo AM, Huang DH, Mozer BA, Dawid IB (1990) The trithorax gene, a trans-acting regulator of the bithorax complex in *Drosophila*, encodes a protein with zinc-binding domains. Proc Natl Acad Sci USA 87: 2112–2116

McCabe NR, Burnett RC, Gill HJ, Thirman MJ, Mbangkollo D, Kipiniak M, van Melle E, Ziemin-van der Poel S, Rowley JD, Diaz MO (1992) Cloning of cDNAs of the *MLL* gene that detect DNA rearrangements and altered RNA transcripts in human leukemic cells with 11q23 translocations. Proc Natl Acad Sci USA 89: 11794–11798

McCabe NR, Kipiniak M, Kobayashi H, Thirman M, Gill H, Rowley JD, Diaz MO (1994) DNA rearrangements and altered transcripts of the *MLL* gene in a human T-ALL cell line Karpas 45 with a t(X;11)(q13;q23) translocation. Genes Chromosom Cancer 9: 221–224

McGinnis W, Krumlauf R (1992) Homeobox genes and axial patterning. Cell 68: 283–302

Milner CM, Campbell RD (1993) The G9a gene in the human major histocompatibility complex encodes a novel protein containing ankyrin-like repeats. Biochem J 290: 811–818

Mitani K, Kanda Y, Ogawa S, Tanaka T, Inazawa J, Yazaki Y, Hirai H (1995) Cloning of several species of *MLL/MEN* chimeric cDNAs in myeloid leukemia with t(11;19)(q23;p13.1) translocation. Blood 85: 2017–2024

Morgan GJ, Cotter F, Katz FE, Ridge SA, Domer P, Korsmeyer S, Wiedemann LM (1992) Breakpoints at 11q23 in infant leukemias with the t(11;19) (q23;p13) are clustered. Blood 80: 2172–2175

Morrissey J, Tkachuk DC, Milatovich A, Francke U, Link M, Cleary ML (1993) A serine/proline-rich protein is fused to HRX in t(4;11) acute leukemias. Blood 81: 1124–1131

Nakamura T, Alder H, Gu Y, Prasad R, Canaani O, Kamada N, Gale RP, Lange B, Crist WM, Nowell PC, Croce CM, Canaani E (1993) Genes on chromosomes 4, 9, and 19 involved in 11q23 abnormalities in acute leukemia share sequence homology and/or common motifs. Proc Natl Acad Sci USA 90: 4631–4635

Naoe T, Kubo K, Kiyoi H, Ohno R (1993) Involvement of the *MLL/ALL-1* gene associated with multiple point mutations of the N-*ras* gene in acute myeloid leukemia with t(11;17)(q23;q25). Blood 82: 2260–2261

Negrini M, Felix CA, Martin C, Lange BJ, Nakamura T, Canaani E, Croce CM (1993) Potential topoisomerase II DNA-binding sites at the breakpoints of a t(9;11) chromosome translocation in acute myeloid leukemia. Cancer Res 53: 4489–4492

Paro R (1990) Imprinting a determined state into the chromatin of *Drosophila*. Trends Genet 6: 416–421

Paro R (1993) Mechanisms of heritable gene repression during development of *Drosophila*. Curr Opin Cell Biol 5: 999–1005

Parry P, Djabali M, Bower M, Khristich J, Waterman M, Gibbons B, Young BD, Evans G (1993) Structure and expression of the human trithorax-like gene 1 involved in acute leukemias. Proc Natl Acad Sci USA 90: 4738–4742

Parry P, Wei Y, Evans G (1994) Cloning and characterization of the t(X;11) breakpoint from a leukemic cell line identify a new member of the forkhead gene family. Genes, Chromosom Cancer 11: 79–84

Pedersen-Bjergaard J, Rowley JD (1994) The balanced and the unbalanced chromosome aberrations of acute myeloid leukemia may develop in different ways and may contribute differently to malignant transformation. Blood 83: 2780–2786

Pocock CFE, Malone M, Booth M, Evans M, Morgan G, Greil J, Cotter FE (1995) BCL-2 expression by leukaemic blasts in a SCID mouse model of biphenotypic leukaemia associated with the t(4;11)(q21;q23) translocation. Brit J Haematol 90: 855–867

Prasad R, Gu Y, Alder H, Nakamura T, Canaani O, Saito H, Huebner K, Gale RP, Nowell PC, Kuriyama K, Miyazaki Y, Croce CM, Canaani E (1993) Cloning of the *ALL-1* fusion partner, the *AF-6* gene, involved in acute myeloid leukemias with the t(6;11) chromosome translocation. Cancer Res 53: 5624–5628

Prasad R, Leshkowitz D, Gu Y, Alder H, Nakamura T, Saito H, Huebner K, Berger R, Croce CM, Canaani E (1994) Leucine-zipper dimerization motif encoded by the *AF17* gene fused to *ALL-1 (MLL)* in acute leukemia. Proc Natl Acad Sci USA 91: 8107–8111

Pui CH, Behm FG, Downing JR, Hancock ML, Shurtleff SA, Ribeiro RC, Head DR, Mahmoud HH, Sandlund JT, Furman WL, Roberts WM, Crist WM, Raimondi SC (1994) 11q23/*MLL* rearrangement confers a poor prognosis in infants with acute lymphoblastic leukemia. J Clin Oncol 12: 909–915

Rastelli L, Chan CS, Pirrotta V (1993) Related chromosome binding sites for *zeste*, suppressor of *zeste* and *Polycomb* group proteins in *Drosophila* and their dependence on *Enhancer of zeste* function. EMBO J 12: 1513–1522

Reeves R (1992) Chromatin changes during the cell cycle. Current Biol 4: 413–423

Reeves R, Nissen MS (1990) The AT-DNA-binding domain of mammalian high mobility group chromosomal proteins: a novel peptide motif for recognizing DNA structure. J Biol Chem 265: 8573–8582
Rowley JD (1992) The der(11) chromosome contains the critical breakpoint junction in the 4;11, 9;11, and 11;19 translocations in acute leukemia. Genes Chromosom Cancer 5: 264–266
Rowley JD, Diaz MO, Espinosa R, Patel YD, van Melle E, Ziemin S, Taillon-Miller P, Lichter P, Evans GA, Kersey JH, Ward DC, Domer PH, LeBeau MM (1990) Mapping chromosome band 11q23 in human acute leukemia with biotinylated probes: identification of 11q23 translocation breakpoints with a yeast artificial chromosome. Proc Natl Acad Sci USA 87: 9358–9362
Rubnitz JE, Link MP, Shuster JJ, Carroll AJ, Hakami N, Frankel LS, Pullen DJ, Cleary ML (1994a) Frequency and prognostic significance of *HRX* rearrangements in infant acute lymphoblastic leukemia: a Pediatric Oncology Group study. Blood 84: 570–573
Rubnitz JE, Morrissey J, Savage PA, Cleary ML (1994b) ENL, the gene fused with HRX in t(11;19) leukemias, encodes a nuclear protein with transcriptional activation potential in lymphoid and myeloid cells. Blood 84: 1747–1752
Sachs AB, Deardorff JA (1992) Translation initiation requires the PAB-dependent poly(A) ribonuclease in yeast. Cell 70: 961–973
Schichman SA, Caligiuri MA, Gu Y, Strout MP, Canaani E, Bloomfield CD, Croce CM (1994a) *All-1* partial duplication in acute leukemia. Proc Natl Acad Sci USA 91: 6236–6239
Schichman SA, Caligiuri MA, Strout MP, Carter SL, Gu Y, Canaani E, Bloomfield CD, Croce CM (1994b) *All-1* tandem duplication in acute myeloid leukemia with a normal karyotype involves homologous recombination between *Alu* elements. Cancer Res 54: 4277–4280
Schichman SA, Canaani E, Croce CM (1995) Self-fusion of the *ALL1* gene. A new genetic mechanism for acute leukemia. JAMA 273: 571–576
Sedkov Y, Tillib S, Mizrokhi L, Mazo A (1994) The bithorax complex is regulated by *trithorax* earlier during *Drosophila* embryogenesis than is the Antennapedia complex, correlating with a bithorax- like expression pattern of distinct early *trithorax* transcripts. Development 120: 1907–1917
Slovak ML, Traweek ST, Willman CL, Head DR, Kopecky KJ, Magenis RE, Appelbaum FR, Forman SJ (1995) Trisomy 11: an association with stem/progenitor cell immunophenotype. Brit J Haematol 90: 266–273
Sorensen PH, Chen CS, Smith FO, Arthur DC, Domer PH, Bernstein ID, Korsmeyer SJ, Hammond GD, Kersey JH (1994) Molecular rearrangements of the *MLL* gene are present in most cases of infant acute myeloid leukemia and are strongly correlated with monocytic or myelomonocytic phenotype. J Clin Invest 93: 429–437
Stock W, Thirman MJ, Dodge RK, Rowley JD, Diaz MO, Wurster-Hill D, Sobol RE, Davey FR, Larson RA, Westbrook CA, Bloomfield CD (1994) Detection of *MLL* gene rearrangements in adult acute lymphoblastic leukemia. A Cancer and Leukemia Group B study. Leukemia 8: 1918–1922
Surowy C, Hoganson G, Gosink J, Strunk K, Spritz R (1990) The human RD protein is closely related to nuclear RNA-binding proteins that has been conserved. Gene 90: 299–302
Susuki M (1989) SPKK, a new nucleic acid-binding unit of protein found in histone. EMBO J 8: 797–804
Tamkun JW, Deuring R, Scott MP, Kissinger M, Pattatucci AM, Kaufman TC, Kennison JA (1992) brahma: A regulator of Drosophila homeotic genes structurally related to the yeast transcriptional activator SNF2/SWI2. Cell 68: 561–572
Theissen J, Etzerodt M, Reuter R, Schneider C, Lottspeich F, Argos P, Lührmann R, Philipson L (1986) Cloning of the human cDNA for the U1 RNA-associated 70K protein. EMBO J 5: 3209–3217
Thirman MJ, Gill HJ, Burnett RC, Mbangkollo D, McCabe NR, Kobayashi H, Ziemin-van der Poel S, Kaneko Y, Morgan R, Sandberg AA, Chaganti RSK, Larson RA, Le Beau MM, Diaz MO, Rowley JD (1993a). Rearrangement of the *MLL* gene in acute lymphoblastic and acute myeloid leukemias with 11q23 chromosomal translocations. New Engl J Med 329: 909–914
Thirman MJ, Mbangkollo D, Kobayashi H, McCabe NR, Gill HJ, Rowley JD, Diaz MO (1993b) Molecular analysis of 3′ deletions of the *MLL* gene in 11q23 translocation reveals that the zinc finger domain of *MLL* are often deleted. Proc American Assoc Cancer Res 34: 495 (abstr)
Thirman MJ, Levitan DA, Kobayashi H, Simon MC, Rowley JD (1994) Cloning of *ELL*, a gene that fuses to *MLL* in a t(11;19)(q23;p13.1) in acute myeloid leukemia. Proc Natl Acad Sci USA 91: 12110–12114.
Thompson KA, Wang B, Argraves WS, Giancotti FG, Scxhrank DP, Ruoslahti E (1994) BR140, a novel zinc-finger protein with homology to the TAF250 subunit of TFIID. Biochem Biophys Res Commun 198: 1143–1152

Tkachuk DC, Kohler S, Cleary ML (1992) Involvement of a homolog of *Drosophila trithorax* by 11q23 chromosomal translocations in acute leukemias. Cell 71: 691–700

Tschiersch B, Hofmann A, Krauss V, Dorn R, Korge G, Reuter G (1994) The protein encoded by the *Drosophila* position-effect variegation suppressor gene *Su(var)3–9* combines domains of antagonistic regulators of homeotic gene complexes. EMBO J 13: 3822–3831

Tsukiyama T, Becker PB, Wu C (1994) ATP-dependent nucleosome disruption at a heat-shock promoter mediated by binding of GAGA transcription factor. Nature 367: 525–532

Tse W, Zhu W, Chen HS, Cohen A (1995) A novel gene, *AF1q*, fused to *MLL* in t(1;11)(q21;q23), is specifically expressed in leukemic and immature hematopoietic cells. Blood 85: 650–656

Welch MD, Drubin DG (1994) A nuclear protein with sequence similarlity to proteins implicated in human leukemias is important for cellular morphogenesis and actin cytoskeletal function in *Saccharomyces cerevisiae*. Mol Biol Cell 5: 617–632

Winston F, Carlson M (1992) Yeast SNF/SWI transcriptional activators and the SPT/SIN chromatin connection. Trends Genet 8: 387–392

Wolfe AP (1991) Developmental regulation of chromatin structure and function. Trends Cell Biol 1: 61–66

Wolfe AP (1994) Transcription: in tune with histones. Cell 77: 13–16

van der Lugt NMT, Domen J, Linders K, van Roon M, Robanus-Maandag E, te Riele H, van der Valk M, Deschamps J, Sofroniew M, van Lohuizen M, Berns A (1994) Posterior transformation, neurological abnormalities, and severe hematopoietic defects in mice with targeted deletion of the *bmi-1* proto-oncogene. Genes Dev 8: 757–769

van Lohuizen M, Frasch M, Wientjes E, Berns A (1991) Sequence similarlity between the mammalian *bmi*-1 proto-oncogene and the *Drosophila* regulatory genes *Psc* and *Su*(*z*)2. Nature 353: 353–355

Yamamoto K, Seto M, Akao Y, Iida S, Nakazawa S, Oshimura M, Takahashi T, Ueda R (1993a) Gene rearrangement and truncated mRNA in cell lines with 11q23 translocation. Oncogene 8: 479–485

Yamamoto K, Seto M, Komatsu H, Iida S, Akao Y, Kojima S, Kodera Y, Nakazawa S, Ariyoshi Y, Takahashi T, Ueda R (1993b) Two distinct portions of LTG19/ENL at 19p13 are involved in t(11;19) leukemia. Oncogene 8: 2617–2625

Yu BD, Hess JL, Horning SE, Brown G, Korsmeyer SJ (1995) The *mixed-lineage leukemia* gene is required for *Hox* gene expression and proper segment identity. Blood 86: 123a

Zeleznik-Le NJ, Harden AM, Rowley JD (1994) 11q23 translocations split the "AT-hook" cruciform DNA-binding region and the transcriptional repression domain from the activation domain of the mixed-lineage leukemia (*MLL*) gene. Proc Natl Acad Sci USA 91: 10610–10614

Ziemen-van der Poel S, McCabe NR, Gill HJ, Espinosa R III, Patel Y, Harden A, Rubinelli P, Smith SD, LeBeau MM, Rowley JD, Diaz MO (1991) Identification of a gene, *MLL*, that spans the breakpoint in 11q23 translocations associated with human leukemias. Proc Natl Acad Sci USA 88: 10735–10739

Zink B, Engstrom Y, Gehring WJ, Paro R (1991) Direct interaction of the Polycomb protein with *Antennapedia* regulatory sequences in polytene chromosomes of *Drosophila melanogaster*. EMBO J 10: 153–162

E2A-Pbx1 Induces Growth, Blocks Differentiation, and Interacts with Other Homeodomain Proteins Regulating Normal Differentiation

M.P. KAMPS

1 Perspective

Arrested differentiation is a hallmark of progenitor cell leukemias as well as many other types of human cancer. It is hypothesized that such progenitor cell cancers express oncoproteins that block differentiation. The E2A-Pbx1 oncoprotein that results from the t(1;19) chromosomal translocation of childhood pre-B cell acute lymphoblastic leukemia (pre-B ALL) is a fascinating oncoprotein because it blocks

Department of Pathology, University of California, San Diego, School of Medicine, 9500 Gilman Drive, La Jolla, CA 92093-0612, USA

differentiation and physically interacts with homeodomain proteins, which are effectors of normal differentiation. These properties suggest the attractive hypothesis that E2A-Pbx1 blocks pre-B cell differentiation by interfering with the ability of other homeodomain proteins to orchestrate terminal B cell differentiation.

2 Discovery of the pre-B ALL Phenotype and the (1;19) (q23;p13.3) Chromosomal Translocation

The pre-B cell phenotype of childhood ALL was first described in 1978 by VOGLER et al. This phenotype was distinctive from others in that the majority of leukemic progenitor cells expressed cytosolic immunoglobulin (cIg) μ heavy chains but lacked expression of cell surface Ig (VOGLER et al. 1978). This same phenotype also typifies normal pre-B cells. Approximately 24% of newly diagnosed pediatric ALL (89 of 369 cases) exhibit the pre-B cell immunophenotype (CARROLL et al. 1984). In 1984, the (1;19)(q23;p13.3) chromosomal translocation was first described in pediatric pre-B cell ALL (WILLIAMS et al. 1984; CRIST et al. 1990). Since then, it has become apparent that 23% of pediatric pre-B ALL (5% of pediatric ALS) contain this translocation (CARROLL et al. 1984), making the t(1;19) the most common translocation in childhood ALL (CRIST et al. 1990), and one of the more common translocations in human leukemia. The t(1;19) is not observed in other leukemias, and children with pre-B ALL containing the t(1;19) have a poorer prognosis for responding to conventional therapy than do children whose pre-B ALL cells lack detectable chromosomal translocations (CRIST et al. 1990). Thus, it has been theorized that an oncogene created by the t(1;19) induces a cellular phenotype that responds poorly to therapy. A more aggressive therapy has now proven more effective on these t(1;19) cases, offsetting the negative prognostic impact of this translocation (RAIMONDI et al. 1990). While some t(1;19) cells contain a balanced translocation, and therefore contain both the derivative 1(1–) and 19(19+) chromosomes, many times the der 1 chromosome is lost, suggesting that the der 19 chromosome contains an oncogene at or near the translocation breakpoint.

3 E2A, Pbx1, and the E2A-Pbx1 Fusion Protein

The (1;19)(q23;p13.3) chromosomal translocation fuses the coding regions of the *E2A* transcription factor gene on chromosome 19 with the novel *PBX1* homeobox gene on chromosome 1, producing the chimeric gene *E2A-PBX1* and the chimeric protein E2A-Pbx1 (Fig. 1A). E2A was first identified as a transcriptional activator that bound to the immunoglobulin κ light chain enhancer. It belongs to the family of transcription factors containing helix-loop-helix DNA-binding domains (MURRE

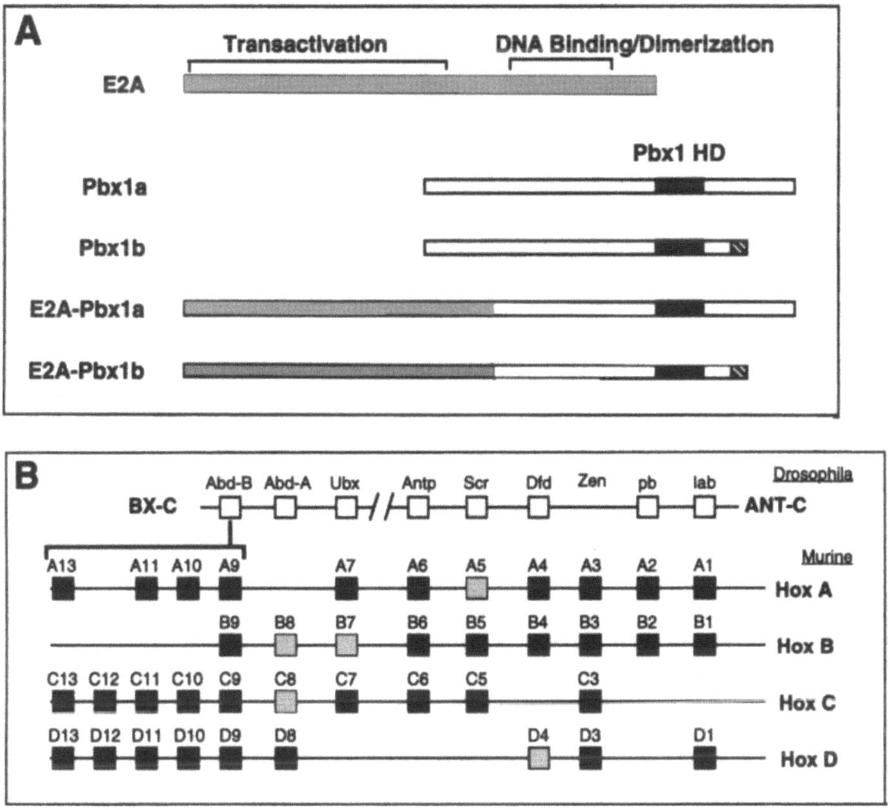

Fig. 1A,B. A Structures of *Pbx1* and *E2A-Pbx1*. *Stippled regions* represent *E2A* sequences and *open regions* represent *Pbx1* sequences. The *Pbx1* homeodomain, indicated by "Pbx1 HD", is *blackened,* and the unique COOH-terminal sequence of *Pbx1b* is designated by a *cross-hatched region*. **B** Comparison of the homebox genes contained in the *Drosophila* HOM-C complex with homologues contained in the four murine *Hox* gene clusters. The *boxes* representing *Hox* genes proven to exhibit cooperative binding with *Pbx1* are *stippled*

et al. 1989), and it is strongly expressed in the B cell lineage. *E2A* was mapped to 19p13.3, the precise location of the chromosome 19 breakpoint of the t(1;19) translocation, and an altered form of the *E2A* transcript was discovered in these t(1;19) cells (Mellentin et al. 1989). This novel transcript encoded a fusion between the upstream *trans*-activation domains of *E2A* and the majority of *PBX1,* including its homeodomain (Kamps et al. 1990; Nourse et al. 1990). Because *PBX1* is not expressed in the B cell lineage, translocation with E2A induces both expression and mutation of Pbx1. With few exceptions (Privitera et al. 1994; Numata et al. 1993) the same fusion point arises in chimeric E2A-Pbx1 mRNAs, resulting from breakpoints that lie within the same introns of E2A and Pbx1 (Hunger et al. 1991; Privitera et al. 1992). Eliminated from the fusion protein are E2A sequences containing the helix-loop-helix, DNA-binding domain (Murre et al. 1989), and the first 89 amino acids of Pbx1. Thus, the structure of E2A-Pbx1 suggests that it may be oncogenic because it activates transcription of Pbx1 target genes. Differential mRNA splicing produces two forms of Pbx1 in normal cells (Pbx1a and Pbx1b) and two forms of E2A-Pbx1 in pre-B ALL cells (E2A-Pbx1a and E2A-Pbx1b; KAMPS et al. 1991; NOURSE et al. 1990).

4 The *PBX* Gene Family and Interspecies Homologues

PBX1 is one member of a family of vertebrate homeobox genes that include *PBX2* and *PBX3* (MONICA et al. 1991). Pbx1 exhibits 94% sequence identity with Pbx2 and Pbx3 between residues 40 and 315, but diverges substantially from Pbx2 and Pbx3 at its NH_2-terminal and COOH-terminal ends. While *PBX1* is expressed in all tissues except the B and T cell lineages, *PBX2* and *PBX3* are expressed ubiquitously (MONICA et al. 1991). The *Drosophila* gene *extradentical* (EXD) is a clear *PBX* homologue, exhibiting over 71% sequence identity to *PBX1* (RAUSKOLB et al. 1993). Likewise, the *ceh-20* gene of *C. elegans* encodes a protein whose homeodomain is 90% identical to that of Pbx1 and in which 63 of 94 residues upstream of the homeodomain are identical to that of Pbx1 (BURGLIN and RUVKUN 1992). Recently, the copper homeostasis gene *CUP9* of yeast was found to encode a homeodomain 47% identical to Pbx1 (KNIGHT et al. 1994). Thus, *PBX* genes have been conserved throughout eukaryotic evolution.

5 Transcriptional Properties of Pbx1 vs E2A-Pbx1

Using recombinant proteins, the homeodomains of Pbx1, Pbx2, and Pbx3 were found to bind the DNA motif ATCAATCAA, which we designated the PRS, for Pbx1-responsive sequence (LEBRUN and CLEARY 1994; LU et al. 1994; VAN DIJK et al. 1993). E2A-Pbx1 strongly activates transcription through the PRS, while Pbx1 does not (LU et al. 1994), suggesting that E2A-Pbx1 causes transformation by activating expression of genes containing a sequence similar to the PRS, which may normally be regulated by Pbx1, Pbx2, or Pbx3.

6 Transforming Abilities of E2A-Pbx1

In animal models, expression of E2A-Pbx1a induces both myeloid and T lymphoid leukemia (KAMPS and BALTIMORE 1993; DEDERA et al. 1993), and expression of either E2A-Pbx1a or E2A-Pbx1b in mouse marrow cultured in the presence of granulocyte-macrophage colony stimulating factor (GM-CSF) results in the rapid outgrowth of immortalized, GM-CSF-dependent, myeloblast cell lines (KAMPS and WRIGHT 1994). This suggests that E2A-Pbx1 can block growth-factor induced differentiation in hematopoietic progenitor cells, therein maintaining the cell in an immature state in which it continues to proliferate in response to lymphokines in its environment. Both E2A-Pbx1a and E2A-Pbx1b also induce tumorigenic conversion of NIH3T3 fibroblasts, as assayed by tumor formation in nude mice (KAMPS et al. 1991). Thus, once a pre-B cell acquires the t(1;19) translocation, E2A-Pbx1 may exert two transforming effects – blocking differentiation and inducing growth. Overexpression of Pbx1 does not induce transformation in any of these models.

Analysis of the importance of the *trans*-activation and DNA-binding functions of E2A-Pbx1 has yielded an unexpected result. Although the presence of sequences possessing *trans*-activation function is essential, the DNA-binding activity of the Pbx1 homeodomain is dispensable for both fibroblast transformation and for generation of T cell leukemia in transgenic mice (MONICA et al. 1994; KAMPS et al. 1995). DNA-binding-independent transformation of fibroblasts has been observed both for mutants of E2A-Pbx1 in which the homeodomain is deleted (MONICA et al. 1994), as well as in point mutants that disrupt homeodomain-dependent DNA-binding (KAMPS et al. 1995). This suggests that E2A-Pbx1 can exhibit a mitogenic function by activating transcription of certain target genes strictly through protein-protein interactions. DNA-binding mutants of E2A-Pbx1 are, however, severely compromised in their ability to block myeloid differentiation in our cultured marrow assay (KAMPS et al. 1995), suggesting that blocking differentiation requires alteration of a different subset of target genes. This observation raises the issue that the T cell leukemias induced by expression of the DNA-binding mutant of E2A-Pbx1 may contain a second oncogene that blocks T cell differentiation. Because the ability of E2A-Pbx1 to block myeloid differentiation requires its DNA-binding function, we would suggest that DNA-binding will also be an essential feature for the full oncogenic role of E2A-Pbx1 in human pre-B ALL.

7 Does Pbx1 Regulate Development?

A variety of evidence suggests that Pbx1 regulates transcription through its interactions with other transcription factors. Distinct but limited sequence homology between the Pbx1 homeodomain and the primitive homeodomain of the yeast repressor protein a1 suggests that Pbx1 may share biochemical features with a1, such as cooperative binding to DNA with other transcription factors (LEVINE and HOEY 1988; SCOTT et al. 1989; WRIGHT et al. 1989). The first indication that Pbx1 was important for the appropriate regulation of normal development arose from identification of the *extradentical* gene (*EXD*) in *Drosophila*, whose mutation causes homeotic transformations by altering the morphological consequences of homeotic selector gene expression. When sequenced *EXD* exhibited 71% identity with *PBX1*

(Rauskolb et al. 1993). In *Drosophila*, expression of homeobox genes of the antennapedia and bithorax complexes (*ANT-C* and *BX-C*, respectively, Fig. 1B), which together are designated the homeotic complex (*HOM-C*), establishes anterior-posterior identity of larval structures (McGinnis and Krumlauf 1992). Genetic evidence suggests that Exd regulates normal differentiation, in part, by functioning in concert with the products of certain HOM-C homeobox genes to activate or repress target gene expression. For instance, both Exd and Ultrabithorax (Ubx) are required to activate expression of the *decapentaplegic* (*DPP*) gene in parasegments posterior to number 7 (Rauskolb and Wieschaus 1994). When Exd was found to bind DNA cooperatively with the Abdominal A (Abd-A) or Ubx homeodomain proteins, but not with Antennapedia or Abdominal B (Abd-B, Van Dijk and Murre 1994; Chan et al. 1994), it suggested that Pbx proteins may likewise contribute to differentiation in vertebrates through their cooperative interactions with Hox proteins on target genes.

8 Cooperative Binding to ATCAATCAA by Vertebrate Hox Proteins and Pbx1 or E2A-Pbx1

In mice and humans, four loci (*HOX-A, HOX-B, HOX-C*, and *HOX-D*; Fig. 1B) contain linear arrays of homeobox genes that are similar to *Drosophila HOM-C* genes in their embryonic expression patterns and in the amino acid sequence of their DNA-binding homeodomains, but not in sequences outside the homeodomain (McGinnis and Krumlauf 1992). In mice, aberrant expression of these *Hox* genes produces homeotic transformations of structures along the anterior-posterior axis, indicating that *Hox* genes, like their *Drosophila* cognates, play a role in establishing differentiation of anterior-posterior structures (Charite et al. 1994; Chisaka and Capecchi 1991; Chisaka et al. 1992; Condie and Capecchi 1993; Dolle et al. 1993; Jeannotte et al. 1993; Jegalian and DeRobertis 1992; LeMouellic et al. 1992; Lufkin et al. 1991, 1992; Morgan et al. 1992; Pollock et al. 1992; Ramirez-Solis et al. 1993). Based on the physical interaction of Exd with Abd-A and Ubx in *Drosophila* and the similar functions of *HOM-C* genes and *HOX* genes, we investigated whether Pbx1 and E2A-Pbx1 exhibit cooperative DNA-binding with specific homeodomain proteins encoded by genes of the four *HOX* loci. Thus far, we have examined three *HOX* genes that are positional homologues of Abd-A and Ubx (*Hox-B7, Hox-B8, Hox-C8*), two that are more closely related to the *Drosophila* homeobox genes Sexcombs reduced (*Scr*) and Deformed (*Dfd*) (*Hox-A5* and *Hox-D4*, respectively), and one homebox gene (*Engrailed-2*) that is not located in the *Hox A*, *B*, *C*, or *D* loci in humans and is unrelated to any of the homeobox genes of the *ANt-C* or *BX-C* in *Drosophila*. Hox-B7, Hox-B8, Hox-C8, and Hox-A5 are also dominant oncoproteins in fibroblasts (Maulbecker and Gruss 1993), and Hox-B8, like E2A-Pbx1, arrests myeloid differentiation in murine marrow cultures (Perkins and Cory 1993; Perkins et al. 1990). Both Pbx1a and Pbx1b, as well as

E2A-Pbx1a and E2A-Pbx1b, cooperated with Hox-A5, Hox-B7, Hox-B8, Hox-C8, and Hox-D4 to form a tight complex on DNA probes containing the PRS itself (Fig. 2). Only the ATCAATCAA sequence is required for these Hox proteins to bind cooperatively with Pbx1, and both factors must retain their ability to bind DNA to form the cooperative-binding complex (Lu et al. 1995). Thus, Pbx1 binds cooperatively to ATCAATCAA with many Hox proteins, suggesting that genetic targets for Pbx1 and E2A-Pbx1 include Hox target genes.

Hox proteins that cooperate with Pbx1 in binding ATCAATCAA, however, fail to cooperate with Pbx1 in binding other optimal Hox-binding sites. For example, mutation of the second C in the ATCAATCAA sequence to T (ATCAATTAA) creates a sequence that binds Hox proteins five times more tightly; however, no Hox protein binds this variant cooperatively with either Pbx1 or E2A-Pbx1. Methylation interference and DNase-protection assays have revealed that Pbx1 is positioned 3′ to the Hox protein in the cooperative-binding complex. Thus, this C residue is likely to be critical for Pbx1-binding. Together, these data suggest

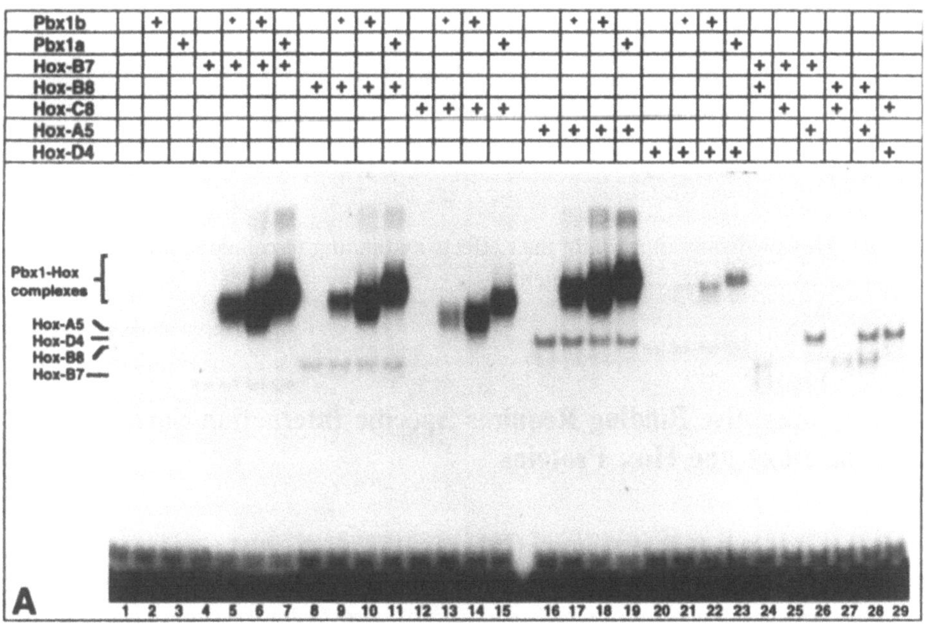

Fig. 2A–B. *Hox-B7, Hox-B8, Hox-C8, Hox-A5,* and *Hox-D4* cooperate with *Pbx1* and *E2A-Pbx1* to form a tight complex on the Pbx1-responsive sequence (PRS) **A** Combinations of in vitro translated Pbx1a or Pbx1b and Hox proteins (as designated above each lane) were incubated with a probe containing the PRS, and protein: DNA complexes were analyzed by native gel electrophoresis. A total of 3 μl of rabbit reticulocyte translation mix was added in all cases. A *plus symbol* represents addition of 1.5 ml of translation mix. The *small plus sign* in the Pbx1b row represents addition of 0.5 ml of translation mix. The amounts of Hox-B7, Hox-B8, Hox-C8, and Hox-A5 added to parallel reactions were equivalent; however, one tenth the amount of Hox-D4 was added due to inefficient expression of Hox-D4 from the vector. **B** Comparison of cooperative binding to the PRS by Hox proteins and E2A-Pbx1. The homeodomain proteins added to each of the binding reactions is indicated above each lane

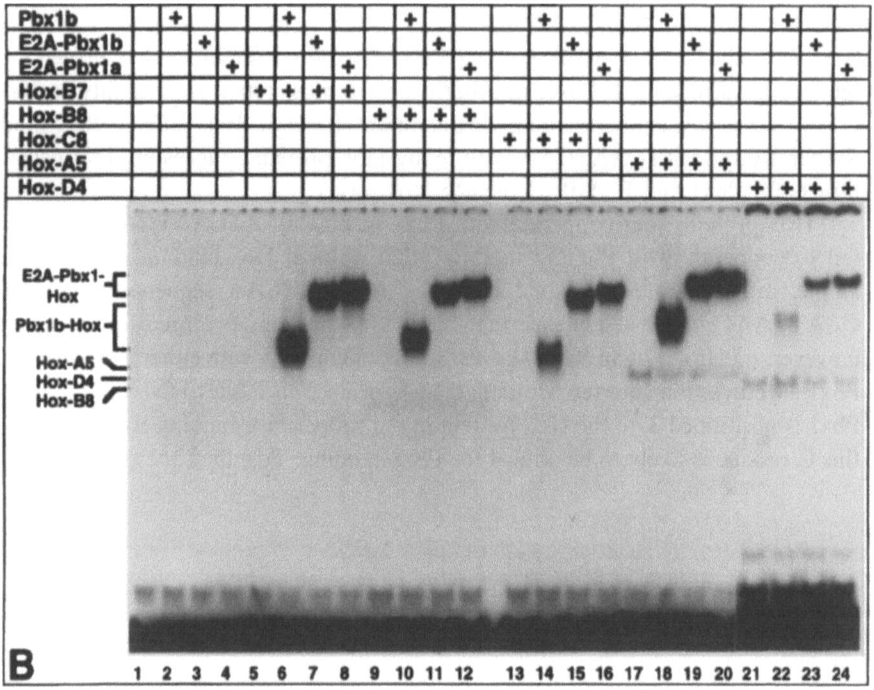

Fig. 2B

that Hox proteins can mediate their effects by binding to cellular promoters in the presence or absence of Pbx proteins.

9 Cooperative Binding Requires Specific Interaction Surfaces on Pbx1 and Hox Proteins

Cooperative binding of Pbx1 and Hox proteins to the PRS suggests that each protein interacts with the other and binds a specific DNA sequence within the PRS. Pbx1 sequences involved in such interactions map to the homeodomain and 20 amino acids on the COOH-terminal side of the homeodomain; however, it is not yet clear which residues within this region are involved in protein-protein interactions (Lu, unpublished observations). By contrast, a highly localized interaction motif is found in Hox proteins. The sequences of Hox genes *A1-A8, B1-B8, C1-C8*, and *D1-D8*, when known, contain a ubiquitous pentapeptide sequence, Y/F-PWM-R/K, located 5–56 amino acids to the NH_2-terminal side of the homeodomain (Fig. 3A). Past studies have found no importance of this motif for the ability of Hox proteins to activate transcription from artificial promoters. By introducing deletions and point mutations into the Hox-B8 and Hox-A5 proteins, we have found that this sequence is essential for cooperative interactions with Pbx1, even though it is absolutely dispensable for monomeric DNA-binding of Hox proteins (KNOEPFLER and KAMPS 1995). An example of such an analysis for Hox-A5, which contains the pentapeptide at residues 177–181, is illustrated in Fig. 3B. NH_2-terminal deletions to residue 147 do not alter the ability of Hox-A5 alone to bind DNA (lanes 4–5) or its ability to bind the PRS cooperatively with Pbx1 (lanes 11–12). While deletion to residue 172 substantially reduces monomeric DNA-binding (lane 6), the protein still exhibits a substantial ability to cooperate with Pbx1 (lane 13). However, while deletion to residue 184 produced a protein that actively bound DNA (lane 7), it was incapable of exhibiting cooperative binding with Pbx1 (lane 14). This deletion encompassed an additional 12 amino acids that also contained the pentapeptide motif. As with Hox-B8, elimination of the COOH terminus of Hox-A5 also reduces DNA-binding affinity (lane 3), but does not alter cooperativity with Pbx1 substantially (lane 10). DNA-binding by the Hox-A5 homeodomain is essential for cooperative binding with Pbx1, because a mutation that converts the invariant asparagine residue at position 51 of the homeodomain to serine (N245S mutant of Hox-A5) abrogates both DNA-binding (lane 8) and cooperativity with Pbx1 (lane 15). Thus, the minimal sequence of Hox-A5 required for cooperative binding with Pbx1 is encompassed by residues 172–256, which contain the homeodomain and pentapeptide motif. Introduction of point mutations within the pentapeptide motif of full-length Hox-A5 reveal that tryptophan-179 of the pentapeptide was essential for cooperative binding with Pbx1 (Fig. 3B, mutant W179A, lane 23) but not for DNA-binding (lane 20). A larger series of point mutations, when introduced within the pentapeptide motif of Hox-B8, revealed that mutation of the tryptophan to either phenylalanine or alanine or mutation of the methionine to either isoleucine or alanine destroyed cooperative binding with Pbx1, while mutation of the proline to alanine had little effect. Such mutational analysis has confirmed the hypothesis that the pentapeptide is dispensable for DNA-binding but essential for cooperative binding with Pbx1. Like Pbx1, E2A-Pbx1 also fails to exhibit cooperative DNA-binding with Hox proteins containing pentapeptide mutations.

Experiments with synthetic peptides containing the pentapeptide motif suggest that the pentapeptide physically binds to Pbx1. The synthetic peptide QPQI<u>YPWMR</u>KLH (PEP-A5^{WT}), which is derived from Hox-A5 and which contains this pentapeptide sequence (underlined), blocks formation of both Hox-Pbx1 and Hox-E2A-Pbx1 complexes on the PRS, while a point mutant version, QPQIYP<u>F</u>MRKLH (PEP-A5^{MUT}), does not (Fig. 3C). This was observed for complexes containing Hox-B7, Hox-B8, and Hox-A5. Mutations at this tryptophan residue also prevent the ability of Hox-A5 to bind to the PRS cooperatively with Pbx1 (Fig. 3B). This observation suggests that a portion of the pentapeptide in the Hox protein binds Pbx1 and that this interaction is competed by addition of the pentapeptide itself, resulting in complex dissociation. Another striking function of peptides containing the pentapeptide sequence is their ability to stimulate Pbx1-mediated DNA-binding. PEP-A5^{WT} increases the ability of Pbx1 to bind DNA at

Protein	Pentapeptide	Distance from HD
Lab	Y K W M Q	109
Hox-A1	F D W M K	19
Hox-B1	F D W M K	19
Hox-D1	F E W M K	19
Pb	Y P W M K	28
Hox-A2	Y P W M K	37
Hox-B2	F P W M K	43
Zen	NONE	
Hox-B3	F P W M K	56
Hox-D3	F P W M K	28
Dfd	Y P W M K	17
Hox-A4	Y P W M K	15
Hox-B4	Y P W M K	15
Hox-C4	Y P W M K	15
Hox-D4	Y P W M K	15
Scr	Y P W M K	14
Hox-A5	Y P W M R	13
Hox-B5	F P W M R	12
Hox-C5	Y P W M T	9
Antp	Y P W M R	7
Hox-B6	Y P W M Q	12
Hox-C6	Y P W M Q	13
Ubx	Y P W M A	12-50
Hox-A7	Y P W M R	5
Hox-B7	Y P W M R	5
AbdA	Y P W M T	24
Hox-B8	F P W M R	5
Hox-C8	F P W M R	4
Hox-D8	F P W M R	5
Others		
STEF	F P W M K	24
Hox-11	F P W M E	21
Hox-7.1	T P W M N	24
Consensus	Y/F P W M K/R	

Fig. 3A–C. The Hox pentapeptide motif is required for cooperative binding with Pbx1. **A** All reported pentapeptide sequences for the *Hox* genes designated 1-8 in Hox loci A-D are indicated below the pentapeptide sequence of their *Drosophila* counterpart. The last category, designated "others", represents similar pentapeptide sequences contained in three genes not contained in the *Hox-A* to *Hox-D* loci. "Distance from the homeodomain" represents the number of amino acids located between the fifth residue of the pentapeptide and the first residue of the homeodomain. In Ubx, differential splicing produces proteins that position the pentapeptide at different distances from the homeodomain

Fig. 3B,C. B Deletion analysis (*lanes* 2–15) and point mutational analysis (*lanes* 16–23) of sequences of Hox-A5 required for DNA-binding in the absence of Pbx1 (*lanes* 2–8, 18–20) and cooperative DNA-binding in the presence of Pbx1 (lanes 9–15, 21–23). All proteins were produced by coupled transcription/translation, using rabbit reticulocyte lysates. Normal and mutant proteins are designated above each lane of the electrophoretic-mobility shift analysis, and the addition of Pbx1 is indicated by plus *sign* in the *top row*. Deletion mutants are referred to by the residues they contain and point mutations are indicated by designation of the wild-type amino acid followed by its position and altered identity. Full-length Hox-A5 is 270 amino acids long, its pentapeptide motif is contained between residues 177 and 181. A picture designating each deletion point is illustrated below the lanes. Approximately equal amounts (less than two fold difference from the average) of each protein were used. Protein abundance was based on incorporation of [^{35}S] methionine, resolution by SDS-PAGE, and quantitation of β-emission. **C** Synthetic peptides containing the pentapeptide sequence disrupt formation of complexes containing Hox proteins and Pbx1 or E2A-Pbx1. Synthetic peptides containing the wild-type or mutant Hox-A5 pentapeptide sequence (*at bottom* of **C**) were added to gel-shift complexes formed by addition of Hox proteins and Pbx1 or E2A-Pbx1. Proteins added to each reaction are indicated by a plus sign above each lane

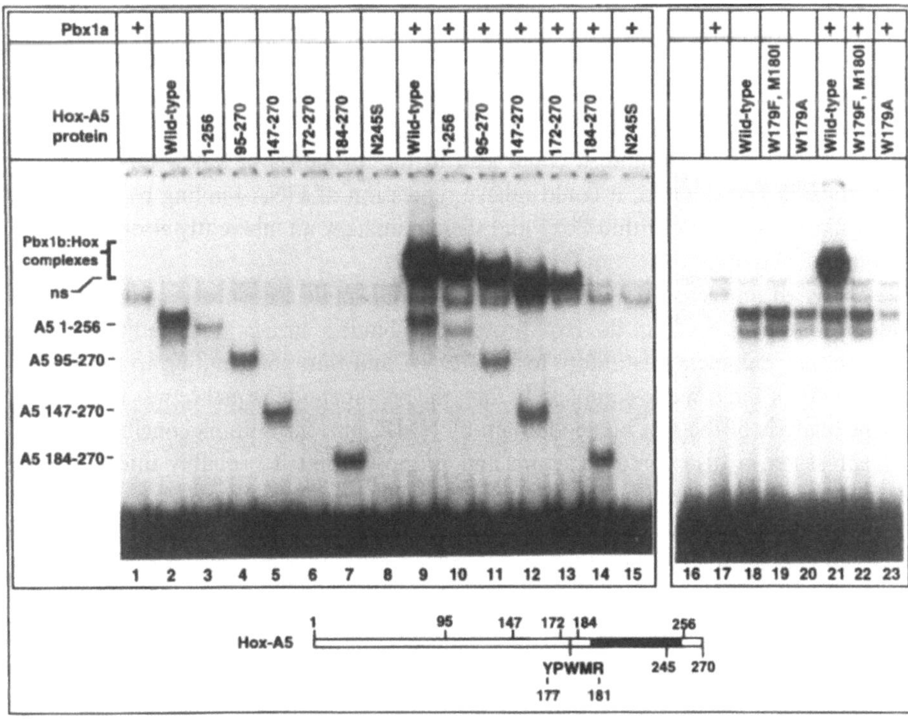
Pbx1a
Hox-A5 protein
Wild-type
1-256
95-270
147-270
172-270
184-270
N245S
W179F, M180I
W179A
Pbx1b:Hox complexes
ns
A5 1-256
A5 95-270
A5 147-270
A5 184-270
1 2 3 4 5 6 7 8 9 10 11 12 13 14 15
16 17 18 19 20 21 22 23
Hox-A5
1 95 147 172 184 256
YPWMR 245 270
177 181

B

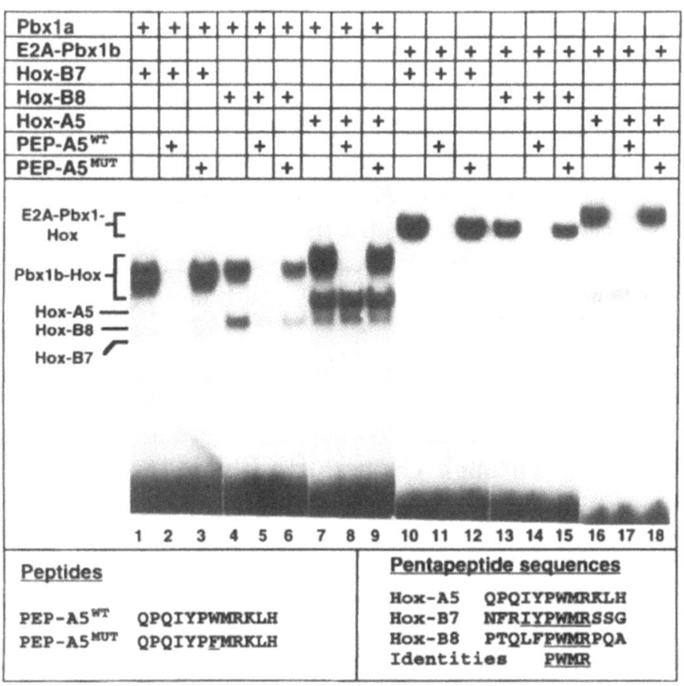
Pbx1a
E2A-Pbx1b
Hox-B7
Hox-B8
Hox-A5
PEP-A5^WT
PEP-A5^MUT
E2A-Pbx1-Hox
Pbx1b-Hox
Hox-A5
Hox-B8
Hox-B7
1 2 3 4 5 6 7 8 9 10 11 12 13 14 15 16 17 18
Peptides
PEP-A5^WT QPQIYPWMRKLH
PEP-A5^MUT QPQIYPFMRKLH
Pentapeptide sequences
Hox-A5 QPQIYPWMRKLH
Hox-B7 NFRIYPWMRSSG
Hox-B8 PTQLFPWMRPQA
Identities PWMR

C

least ten fold while PEP-A5^{MUT} does not, indicating that the tryptophan residue is essential for protein-protein interaction and for increasing the affinity of Pbx1 for DNA KNOEPFLER and KAMPS 1995. The binding of the pentapeptide sequence to Pbx1 could enhance Pbx1 DNA-binding through several mechanisms. First, it could directly contact the Pbx1 homeodomain and stabilize the homeodomain-DNA interaction. Second, it could lead to the formation of a new DNA-binding surface in Pbx1. Third, it could relieve repression of DNA-binding by a negative regulatory structure within Pbx1 and therein unmask an inherently stronger DNA-binding activity.

Based on these data, we envision the pentapeptide sequence as a short protein loop that projects from the Hox protein and binds a surface of the adjacent Pbx1 molecule, enhancing its ability to bind DNA and thus contributing to the overall stability of the trimeric complex. In fact, the pentapeptide sequence was included in the analysis of the Ant homeodomain by NMR, and the authors concluded that it represented a disordered structure and proposed that it probably interacts with another transcription factor (QIAN et al. 1992). In different Hox proteins, the pentapeptide sequence is located at various distances from the homeodomain, and it is tempting to speculate that different Hox proteins will complex with Pbx1 on different DNA targets based, in part, on the distance between this pentapeptide and the homeodomain. In addition to the genes of the *Hox* loci, four other homeodomain transcription factors contain the pentapeptide core sequence upstream of their homeodomain: *STF-1*, which regulates insulin and somatostatin gene expression in the endocrine pancreas; *Hox-11*, which is transcriptionally activated as a consequence of chromosomal translocation in human T cell ALL; and *Msx-1* and *Msx-2*, which are implicated in developmental regulation of multiple structures. Indeed, we find that *STF-1* does bind the PRS cooperatively with Pbx1 and E2A-Pbx1, and the complex containing *STF-1* and Pbx1 or E2A-Pbx1 can be dissociated by inclusion of PEP-A5^{WT}. We have not yet tested these other proteins for their ability to bind the PRS cooperatively with Pbx1. Thus, homeodomain proteins containing this core sequence are likely to regulate transcription with Pbx in vivo. Hox proteins whose designation is numerically greater than 9 in Fig. 1B have homeodomains that are more homologous to that of Abd-B of *Drosophila*, and they do not contain the pentapeptide motif.

10 Generating Diversity Through Cooperative Binding of Hox Proteins and Pbx1

One explanation to account for how individual Hox proteins have specific developmental functions yet contain homeodomains that bind very similar DNA sequences is to suggest that association with other transcription factors provides target gene specificity. The model that Hox proteins bind Pbx1 through their pentapeptide sequences and therein enhance DNA-binding by Pbx1 suggests four mechanisms that could generate selectivity in DNA-binding among a group of possible Pbx1-Hox heterodimers. First, the specificity of amino acids surrounding the pentapeptide of each Hox protein might provide a mechanism to alter both the affinity of the sequence for Pbx1 as well as its ability to increase DNA-binding by Pbx1. This could provide a selective mechanism whereby different Pbx1-Hox heterodimers could be targeted to different subsets of DNA-recognition motifs. Heterodimers providing the strongest overall affinity could be targeted not only to high-affinity sites, but also to lower-affinity sites, while heterodimers providing a weak affinity could bind only high-affinity sites. Second, the interaction of the pentapeptide with Pbx1 could induce amino acids located between the homeodomain and the pentapeptide to interact with DNA, therein extending the zone of DNA contact with the Hox protein. An analysis of optimal DNA-binding sites for different Pbx1-Hox heterodimers should reveal whether cooperation with Pbx1 alters target specificity. If it does, it will be possible to identify Hox sequences that are responsible for generating such specificity. Third, differential spacing between the pentapeptide and the homeodomain could result in changing the optimal spacing between the homeodomains of Pbx1 and Hox protein heterodimers on DNA, and thus varying the sequence of optimal DNA recognition targets. In addition to the large degree of variability in the distance between the pentapeptide and the homeodomain among different Hox proteins (Fig. 3A), individual homeodomain proteins, such as *Drosophila* Ubx, exhibit alternative mRNA splicing that results in the production of four different protcins that position the pentapeptide motif at different distances from the homeodomain. The result of such differential spacing is currently unknown, but will now be able to be addressed in terms of its effect on cooperative DNA-binding with Pbx1 and alteration of target site specificity. Finally, it is possible that, upon association with Pbx1, DNA-binding of a subset of Hox proteins is also elevated, resulting in preferential DNA-binding of this subset of Hox-Pbx1 heterodimers.

11 The Geometric Problem

It is not clear how two different homeodomain proteins geometrically fit on a DNA element as short as the PRS. In the crystal structure of yeast $\alpha 2$, bound to its DNA motif, the third α helix of the homeodomain, which is termed the "recognition helix," contacts nucleotides 1–3 in the major groove and extends contacts to nucleotides 4–7 on the minor grove (Wolberger et al. 1991). Thus, one possibility is that each recognition helix of the two homeodomains are bound in the same orientation adjacent to each other. In this configuration, the upstream homeodomain would contact the minor grove just opposite the recognition helix of the downstream homeodomain. This overlapping configuration could be constrained within 11 nucleotides. Alternatively, the homeodomains could be positioned in opposite orientations, occupying both faces of the DNA. In this case, only nine nucleotides would be required to form a binding site. Determination of the specific sequences contacted by both Pbx1 and Hox proteins may become more evident upon identification of an optimum site for cooperative binding. Distinguishing the exact geometry of Pbx and Hox proteins bound to the PRS will certainly yield new insights into the mechanisms of interaction between DNA and homeodomain proteins.

12 Transcriptional Effects of Coexpression of Pbx1 and Hox Proteins

Cooperative DNA binding between Pbx1 and Hox proteins in vitro suggests that Pbx1 may also cooperate with Hox proteins in vivo to regulate target gene expression and therein contribute to normal differentiation in vertebrates. To investigate whether Pbx1 and Hox proteins can cooperate in vivo to influence transcription, we have performed cotransfection assays using expression constructs encoding Pbx1 proteins and Hox-A5, Hox-B8, or Hox-C8. Hox-B8, Hox-C8 and Hox-D8 can suppress transcription activated by other Hox proteins (Jones et al. 1992; Zappavigna et al. 1994). The ability of Hox-D8 to repress transcription activated by Hox-D9 is DNA-binding independent, requiring only an NH_2-terminal effector domain and helix I of the Hox-D8 homeodomain (Zappavigna et al. 1994). Because helix I of the Hox-D8 homeodomain is also required for tight binding to Hox-D9 in vitro, it was suggested that repression by Hox-D8 is mediated by direct interaction with Hox-D9. In our assays, trans-activation of a PRS-CAT construct by E2A-Pbx1 was repressed by Hox-A5 in a pentapeptide-dependent manner, suggesting that E2A-Pbx1 can interact with Hox proteins in vivo and thus may alter transcriptional function of Hox proteins on target genes.

13 Implications of Pbx1-Hox Cooperativity for Transformation by E2A-Pbx1

When marrow-derived myeloblasts are cultured in GM-CSF, they exhibit concurrent growth and differentiation and ultimately differentiate into nonmitotic macrophages and neutrophils within 4 weeks (Kamps and Wright 1994). However, when these cultures are infected with E2A-Pbx1 virus, their terminal differentiation is blocked, and they proliferate continuously as GM-CSF-dependent progenitors (Kamps and Wright 1994). Because E2A-Pbx1 exhibits cooperative binding with Hox proteins, E2A-Pbx1 could abrogate normal differentiation by replacing Pbx1 or other Pbx proteins in Pbx-Hox regulatory complexes, thus inducing constitutive transcription of specific Hox target genes. This model would be consistent with the facts that pre-B ALL cells express members of the Hox B locus, including Hox-B7 (Mathews et al. 1991; Petrini et al. 1992), and that transcriptional activation by

E2A correlates with the ability of E2A-Pbx1 to block myeloid differentiation (Kamps et al. 1995).

14 Implication of Hox-Pbx1 Cooperativity for the Mechanism of Transformation by Hox Proteins

Since Hox-A1, Hox-A5, Hox-A7, Hox-B7, and Hox-C8 induce malignant conversion of NIH3T3 cells (MAULBECKER and GRUSS 1993), the association between Pbx1 and these Hox proteins is an association between protooncoproteins. This observation may lead to the elucidation of a common mechanism through which both E2A-Pbx1 and Hox proteins induce transformation. Similar to E2A-Pbx, expression of Hox-B8 in marrow cultures produces factor-dependent outgrowths of myeloid progenitors (PERKINS and CORY 1993) and contributes to myeloid leukemia in mouse bone marrow reconstitution experiments (PERKINS et al. 1990). While E2A-Pbx1 might physically replace a normal Pbx protein in a Hox complex, leading to transcriptional activation of a target gene, sufficient overexpression of a transforming Hox protein could substitute for multiple Hox proteins that regulate gene transcription in conjunction with Pbx1. In this manner, both E2A-Pbx1 and oncogenic Hox proteins could alter transcription of the same or overlapping subsets of target genes and produce transformation by a similar mechanism. Examination of this model must await identification of the target genes of E2A-Pbx1 and oncogenic Hox proteins.

15 Implications of Cooperativity with Hox Proteins for the Normal Developmental Role of Pbx1

In addition to suggesting an attractive mechanism for transformation by E2A-Pbx1, cooperative DNA-binding with Hox proteins also suggests a profoundly important role for Pbx1 in the regulation of normal differentiation. Because axial differentiation in *Drosophila* requires the combined functions of *Exd* and certain homeobox genes of the *HOM-C*, Pbx1 may likewise be essential for regulating normal vertebrate development through its cooperative binding with other homeodomain proteins. Mutations of *Hox* genes in mice that generate either gain of function (BALLING et al. 1989; JEGALIAN and DEROBERTIS 1992; LUFKIN et al. 1992; MORGAN et al. 1992; POLLOCK et al. 1992) or loss of function (CHISAKA and CAPECCHI 1991; CHISAKA et al. 1992, CONDIE and CAPECCHI 1993; DOLLE et al. 1993; JEANNOTTE et al. 1993; LEMOUELLIC et al. 1992; LUFKIN et al. 1991; RAMIREZ-SOLIS et al. 1993) reveal that *Hox* genes, like their *Drosophila* counterparts, direct regional embryonic development and are involved in anterior-posterior axial pattern formation. For instance, the ectopic expression of Hox-A7 induces conversion from the normal seven cervical vertebrae to eight cervical vertebrae and is accompanied by variations in the most anterior vertebrae that suggest a posterior to anterior transformation (Kessel et al. 1990). In addition, ectopic expression of Hox-B8 causes duplication of forelimb structures and homeotic transformation of axial structures (Charite et al. 1994), and Hox-A5 is also essential for appropriate axial differentiation (Jeannotte et al. 1993). Although a direct mouse target gene for coregulation by both Pbx1 plus a Hox protein has not yet been identified, the ability of Pbx1 to associate with Hox proteins, including both Hox-B8 and Hox-A5, suggests that Pbx1 and Hox proteins will likely cooperate to regulate normal transcription of certain Hox target genes. It is not likely that all Hox target genes will be coregulated by Pbx proteins because optimal Hox sites, such as TCAAT-TAATT, bind Hox proteins strongly, but fail to exhibit cooperative binding with Pbx1 (Lu et al. 1995). Thus far, the bovine *CYP17* gene is the only known gene to be transcriptionally regulated through a site that binds Pbx1 (Kagawa et al. 1994; Lund et al. 1990).

Cooperative interactions between Pbx1 and Hox proteins represent a form of interaction that was first predicted by sequence analysis of the Pbx1 homeodomain. When initially cloned, Pbx1 was found to contain a divergent homeodomain that was most homologous to that of the distantly related homeodomain yeast a1 (Kamps et al. 1990). Approximatley one third of residues are identical between the homeodomains of Pbx1 and this yeast protein, including a stretch of 10 of 11 amino acids in the second helix of the homeodomain. a1 binds promoter elements cooperatively with α2 in diploid cells, therein repressing haploid-specific genes (Johnson 1992). In haploid cells, α2 binds DNA cooperatively with MCM1, repressing transcription of a-specific genes (Johnson 1992). Pbx and Hox proteins of vertebrates exhibit some functional homologies to the yeast a1 and α2 homeodomain proteins. Interestingly, the core of the DNA sequence recognised by a1 is CATCA, and a similar sequence, ATCA, is contained within the PRS recognition motif of Pbx1. In addition, similar to the pentapeptide sequence of many Hox proteins, yeast α2 contains a flexible NH2-terminal extension of its homeodomain that mediates cooperative binding with MCM1 (Vershon and Johnson 1993). An amino acid sequence on the COOH-terminal side of the α2 homeodomain is involved in contacting a1 (Mak and Johnson 1993). Thus a1 and α2 may represent early progenitors of a class of homeodomain proteins whose interactions with both DNA and other transcription factors control simple switches in differentiation programs. Pbx and Hox proteins may function in an analogous manner in vertebrates, maintaining both DNA-binding and protein-protein interactions with other transcription factors as key elements that target their transcriptional activities to specific genes.

Acknowledgements. This work was supported by NIH grants CA56876 and CA50528.

References

Balling R, Mutter G, Gruss P, Kessel M (1989) Craniofacial abnormalities induced by ectopic expression of the homeobox gene Hox-1.1 in transgenic mice. Cell 58: 337–348

Burglin TR, Ruvkun G (1992) New motif in PBX genes. Nature genetics 1: 319–320

Carroll A, Crist W, Parmley R, Roper M, Cooper M, Finley W (1984) Pre-B cell leukemia associated with chromosome translocation 1;19. Blood 63: 721–724

Chan S-K, Jaffe L, Capovilla M, Botas J, Mann R (1994) The DNA binding specificity of Ultrabithorax is modulated by cooperative interactions with extradenticle, another homeoprotein. Cell 78: 603–615

Charite J, Graaff W, Shen S, Deschamps J (1994) Ectopic expression of Hoxb-8 causes duplication of the ZPA in the forelimb and homeotic transformation of axial structures. Cell 78: 589–601

Chisaka O, Capecchi M (1991) Regionally restricted developmental defects resulting from targeted disruption of the mouse homeobox gene hox-1.5. Nature 350: 473–479

Chisaka O, Musci T, Capecchi M (1992) Developmental defects of the ear, cranial nerves and hindbrain resulting from targeted disruption of the mouse homeobox gene Hox-1.6. Nature 355: 516–520

Condie B, Capecchi M (1993) Mice homozygous for a targeted disruption of Hoxd-3 (Hox-4.1) exhibit anterior transformation of the first and second cervical vertebrae, the atlas and axis. Development 110: 579–595

Crist W, Carroll A, Shuster J, Behm F, Whitehead M, Vietti T, Look A, Mahoney D, Ragab A, Pullen D, Land V (1990) Poor prognosis of children with pre-B acute lymphoblastic leukemia is associated with the t(1;19)(q23;p13): A pediatric oncology group study. Blood 76: 117–122

Dedera DA, Waller EK, Lebrun DP, Sen-Majumdar A, Stevens ME, Barsh GS, Cleary ML (1993) Chimeric homeobox gene E2A-Pbx1 induces proliferation, apoptosis, and malignant lymphomas in transgenic mice. Cell 74: 833–843

Dolle P, Dierich A, LeMeur M, Schimmang T, Schuhbaur B, Chambon P, Duboule D (1993) Disruption of the Hoxd-13 gene induces localized heterochrony leading to mice with neotenic limbs. Cell 75: 431–441

Henthorn P, Kiledijan M, Kadesch T (1990) Two distinct transcription factors that binds the immunoglobulin enhancer mE5/kE2 motif. Science 247: 467–470

Hunger SP, Galili N, Carroll AJ, Crist WM, Link MP, Cleary ML (1991) The t(1;19)(q23;p13) results in consistent fusion of E2A and PBX1 coding sequences in acute lymphoblastic leukemias. Blood 77: 687–693

Jeannotte L, Lemieux M, Charron J, Poirier F, Robertson E (1993) Specification of axial identity in the mouse: role of the Hoxa-5(Hox1.3) gene. Genes Dev 7: 2085–2096

Jegalian B, De Robertis E (1992) Homeotic transformations in the mouse induced by overexpression of a human Hox3.3 transgene. Cell 71: 901–910

Johnson A (1992) A combinatorial regulatory circuit in budding yeast. Transcriptional Regulation 2: 975–1007

Jones F, Prediger E, Bittner D, DeRobertis E, Edelman G (1992) Cell adhesion molecules as targets for Hox genes: Neural cell adhesion molecule promoter activity is modulated by cotransfection with Hox-2.5 and -2.4. Proc Natl Acad Sci USA 89:2086–2090

Kagawa N, Ogo A, Takahashi Y, Iwamatsu A, Waterman MR (1994) A cAMP-regulatory sequence (CRS1) of CYP17 is a cellular target for the homeodomain protein Pbx1. J Biol Chem 269: 18716–18719

Kamps MP, Murre C, Sun X, Baltimore D (1990) A new homeobox gene contributes the DNA binding domain of the t(1;19) translocation proteins in pre-B ALL. Cell 60: 547–555

Kamps MP, Look AT, Baltimore D (1991) The human t(1;19) translocation in pre-B ALL produces multiples nuclear E2A-Pbx1 fusion proteins with differing transforming potentials. Genes Dev 5: 358–368

Kamps MP, Baltimore D (1993) E2A-PBX1, the t(1;19) translocation protein of human pre-B cell acute lymphocytic leukemia, causes acute myeloid leukemia in mice. Mol Cell Biol 13: 351–357

Kamps MP, Wright DD (1994) Oncoprotein E2A-Pbx1 immortalizes cultured myeloid progenitors without abrogating their factor-dependence. Oncogene 9: 3159–3166

Kamps MP, Wright DD, Lu Q (1995) DNA-binding by oncoprotein E2A-Pbx1 is important for blocking differentiation but dispensable for fibroblast transformation. Oncogene 12: 19–30

Kessel M, Balling R, Gruss P (1990) Variations of cervical vertebrae after expression of a Hox1.1 transgene in mice. Cell 61: 301–308

Knight S, Tamai K, Kosman D, Thiele D (1994) Identification and analysis of a saccharomyces cerevisiae copper homeostasis gene encoding a homeodomain protein. Mol Cell Biol 14: 7792–7804

Knoepfler PS, Kamps MP (1995) The pentapeptide motif of Hox proteins is required for cooperative DNA-binding with Pbx1, physically contacts Pbx1 and enhances DNA-binding by Pbx1. Mol Cell Biol 15: 5811–5819

Lai J-S, Herr W (1992) Ethidium bromide provides a simple tool for identifying genuine DNA-independent protein associations. Proc Natl Acad Sci USA 89: 6958–6962

Lebrun DP, Cleary ML (1994) Fusion with E2A alters the transcriptional properties of the homeodomain protein PBX1 in t(1;19) leukemias. Oncogene 9: 1641–1647

LeMouellic H, Lallemand V, Brulet P (1992) Homeosis in the mouse induced by a null mutation in the Hox-3.1 gene. Cell 69: 251–264

Levine M, Hoey T (1988) Homeobox proteins as a sequence-specific transcription factors. Cell 55: 537–540

Lu Q, Wright DD, Kamps MP (1994) Fusion with E2A converts the Pbx1 homeodomain protein into a constitutive transcriptional activator in human leukemias carrying the t(1;19) translocation. Mol Cell Biol 6: 3938–3948

Lu Q, Knoepfler P, Scheele J, Wright DD, Kamps MP (1995) Both Pbx1 and E2A-Pbx1 bind the DNA motif ATCAATCAA cooperatively with the products of multiple murine Hox genes, some of which are themselves oncogenes. Mol Cell Biol 15: 3786–3795

Lufkin T, Dierich A, LeMeur M, Mark M, Chambon P (1991) Disruption of the Hox-1.6 homeobox gene results in defects in a region corresponding to its rostral domain of expression. Cell 66: 1105–1119

Lufkin T, Mark M, Hart C, Dolle P, LeMeur M, Chambon P (1992) Homeotic transformation of the occipital bones of the skull by ectopic expression of a homeobox gene. Nature 359: 835–841

Lund J, Ahlgren R, Wu D, Kagimoto M, Simpson E, Waterman M (1990) Transcriptional regulation of the bovine CYP17 ($P\text{-}450_{17a}$) gene. J Biol Chem 265: 3304–3312

Mak A, Johnson AD (1993) The carboxyl-terminal tail of the homeodomain protein α2 is required for function with a second homeodomain protein. Genes Dev 7: 1862–1870

Mathews C, Detmer K, Boncinelli E, Lawrence H, Largmen C (1991) Erythroid-restricted expression of homeobox genes of the human HOX 2 locus. Blood 78: 2248–2252

McGinnis W, Krumlauf R (1992) Homeobox genes and axial patterning. Cell 68: 283–302

Maulbecker CC, Gruss P (1993) The oncogenic potential of deregulated homeobox genes. Cell Growth and Dif 4: 431–441

Mellentin JD, Murre C, Donlon TA, McCaw PS, Smith S, Caroll A, McDonald M, Baltimore D, Cleary M (1989) The gene for enhancer binding proteins E12/E47 lies at the t(1;19) breakpoint in acute leukemias. Science 246: 379–382

Monica K, Galili N, Nourse J, Saltman D, Cleary ML (1991) PBX2 and PBX3, new homeobox genes with extensive homology to the human proto-oncogene PBX1. Mol Cell Biol 11: 6149–6157

Monica K, LeBrun DP, Dedera D, Brown R, Cleary ML (1994) Transformation properties of the E2A-Pbx1 chimeric oncoprotein: Fusion with E2A is essential, but the Pbx1 homeodomain is dispensable. Mol Cell Biol 14: 8304–8314

Morgan B, Izpisua-Belmonte J, Duboule D, Tabin C (1992) Targeted misexpression of Hox-4.6 in the avian limb bud causes apparent homeotic transformations. Nature 358: 236–239

Murre C, McCaw P, Baltimore D (1989) A new DNA binding and dimerization motif in immunoglobulin enhancer binding, daughterless, MyoD, and Myc proteins. Cell 56: 777–783

Nourse J, Mellentin JD, Galili N, Wilkinson J, Stanbridge E, Smith SD, Cleary ML (1990) Chromosomal translocation t(1;19) results in synthesis of a homeobox fusion mRNA that codes for a potential chimeric transcription factor. Cell 60: 535–546

Numata S, Kato K, Horibe K (1993) New E2A-Pbx1 fusion transcript in a patient with t(1;19)(q23;p13) acute lymphoblastic leukemia. Leukemia 7: 1441–1444

Perkins A, Cory S (1993) Conditional immortalization of mouse myelomonocytic, megakaryocytic and mast cell progenitors by the Hox-2.4 homeobox gene. EMBO J 12: 3835–3846

Perkins A, Kongsuwan K, Visvader J, Adams JM, Cory S (1990) Homeobox gene expression plus autocrine growth factor production elicits myeloid leukemia. Proc Natl Acad Sci 87: 8398–8402

Petrini M, Quaranta M, Testa U, Samoggia P, Tritarelli E, Care A, Cianetti L, Valtieri M, Barletta C, Peschle C (1992) Expression of selected human HOX-2 genes in B/T acute lymphoid leukemia and interleukin-2/interleukin-1 b-stimulated natural killer lymphocytes. Blood 80: 185–193

Pollock R, Jay G, Bieberich C (1992) Altering the boundaries of Hox3.1 expression: evidence for antipodal gene regulation. Cell 71: 911–923

Privitera E, Kamps MP, Hayashi Y, Inaba T, Shapiro LH, Raimondi SC, Behm F, Hendershot L, Carroll AJ, Baltimore D, Look AT (1992) Different molecular consequences of the 1;19 chromosomal translocations in childhood B-cell precursor acute lumphoblastic leukemia. Blood 79: 1781–1788

Privitera E, Luciano A, Ronchetti D, Arico M, Santostasi T, Basso G, Biondi A (1994) Molecular variants of the 1;19 chromosomal translocation in pediatric acute lymphoblastic leukemia (ALL). Leukemia 8: 554–559

Qian Y, Otting G, Furukubo-Tokunaga K, Affolter M, Gehring W, Wuthrich K (1992) NMR structure determination reveals that the homeodomain is connected through a flexible linker to the main body in the Drosophila Antennapedia protein. Proc Natl Acad Sci USA 89: 10738–10742

Raimondi S, Behm F, Roberson P, Williams D, Pui C, Crist W, Look A, Rivera G (1990) Cytogenetics of pre-B acute lymphoblastic leukemia with emphasis on prognostic implications of the t(1;19). J Clin Oncol 8: 1380–1390

Ramirez-Solis R, Zheng H, Whiting J, Krumlauf R, Bradley A (1993) Hoxb-4 (Hox-2.6) mutant mice show homeotic transformation of a cervical vertebra and defects in the closure of the sternal rudiments. Cell 73: 279–294

Rauskolb C, Peifer M, Wieschaus E (1993) Extradenticle, a regulator of homeotic gene activity, is a homologue of the homeobox-containing human proto-oncogene Pbx1. Cell 74: 1101–1112

Rauskolb C, Wieschaus E (1994) Coordinate regulation of downstream genes by extradenticle and the homeotic selector proteins. EMBO J 13: 3561–3569

Scott MP, Tamkun JW, Hartzel GW III (1989) The structure and function of the homeodomain. Biochim Biophys Acta 989: 25–48

Van Dijk MA, Voorhoeve P, Murre C (1993) Pbx1 is converted into a transcriptional activator upon acquiring the N-terminal region of E2A in pre-B cell acute lymphoblastoid leukemia. Proc Natl Acad Sci USA 90: 6061–6065

Van Dijk MA, Murre C (1994) *Extradenticle* raises the DNA-binding specificity of homeotic selector gene products. Cell 78: 617–624

Vershon AK, Johnson AD (1993) A short, disordered protein region mediates interactions between the homeodomain of the yeast α2 protein and the MCM1 protein. Cell 72: 105–112

Vogler L, Crist W, Bockman D, Pearl E, Lawton A, Cooper M (1978) A new phenotype of childhood lymphoblastic leukemia. N Eng J Med 298: 872–878

Williams D, Look A, Melvin S, Roberson P, Dahl G, Flake T, Stass S (1984) New chromosomal translocations correlate with specific immunophenotypes of childhood acute lymphoblastic leukemia. Cell 36: 101–109

Wolberger C, Vershon A, Liu B, Johnson A, Pabo C (1991) Crystal structure of a MATa2 homeodomain-operator complex suggests a general model for homeodomain-DNA interactions. Cell 67: 517–528

Wright CVE, Cho KWY, Oliver G, DeRobertis EM (1989) Vertebrate homeodomain proteins: families of region-specific transcription factors. Trends Biochem Sci 14: 52–56

Zappavigna V, Sartori D, Mavilio F (1994) Specificity of HOX protein function depends on DNA-protein and protein-protein interactions, both mediated by the homeo domain. Genes Dev 8: 732–744

E2A-HLF Chimeric Transcription Factors in Pro-B Cell Acute Lymphoblastic Leukemia

A.T. Look

1 Introduction

Somatically acquired chromosomal translocations result in the abnormal expression of proteins that may contribute to neoplasia by interfering with or mimicking the action of growth factors and their receptors, signal transducers, or nuclear regulatory proteins and transcription factors, which affect gene expression directly. Oncogenesis mediated by transcription factors is particularly important in the acute leukemias and sarcomas, malignancies in which chromosomal translocations and inversions commonly activate genes encoding transcriptional regulatory proteins (Rowley et al. 1993; Rabbitts 1994). The modular organization of transcription factors provides an ideal mechanism for mediating their multiple effects on cell lineage-specific gene expression: binding to DNA, *trans*-activation or *trans*-repression of target genes and protein-protein interactions within complex regulatory networks.

The modular structure of these factors also permits recombination of functional regions between two such genes by chromosomal rearrangement, forming in the process hybrid transcription factors unique in nature to the tumorigenic clones harboring these chromosomal abnormalities. Two such chimeric oncoproteins are formed with the E2A protein in leukemias arising in early B lineage progenitors. The *E2A-PBX1* chimera, resulting from a t(1;19)(q23;p13) chromosomal translo-

Department of Experimental Oncology, St. Jude Children's Research Hospital, 332 North Lauderdale, Memphis, TN 38105, USA

cation is perhaps the best known, affecting about one fourth of leukemic blasts that are arrested at the pre-B cell stage of development, indicated by the fact that they express cytoplasmic but not surface immunoglobulin heavy chains. In this rearrangement, the *E2A* gene on chromosome 19, which encodes a basic-helix-loop-helix (bHLH) transcription factor, is fused to a homeobox gene (*PBX1*) on chromosome 1, leading to the production of several different forms of hybrid *E2A-PBX1*, oncoproteins (Kamps et al. 1990; Nourse et al. 1990; Izraeli et al. 1992; Numata et al. 1993). These hybrid proteins retain only the NH_2-terminal *trans*-activation domain of E2A; its bHLH domain is absent, replaced by the homeobox DNA-binding and protein interaction domain of PBX1. This suggests that the gene targets of the chimeric E2A-PBX1 protein are recognized by the homeodomain of PBX1, whose normal tissue distribution does not include lymphoid or other hematopoietic cells. In agreement with this interpretation, recent studies have demonstrated that PBX1 is a sequence-specific DNA-binding protein which is converted into a positive transcriptional regulator upon acquisition of the *trans*-activation domain of E2A (Van Dijk et al. 1993; Lu et al. 1994; LeBrun and Cleary 1994). The transforming potential and biochemistry of E2A-PBX1 proteins are the subject of another chapter in this volume.

This review will focus on a second fusion gene in ALL produced by a t(17;19)(q22;p13) chromosomal translocation that incorporates the same NH_2-terminal sequences of *E2A* (including the *trans*-activation domain) as *E2A-PBX1*, in this case joined to the DNA-binding and protein dimerization regions of a previously unidentified member of the bZIP superfamily of transcription factors, which has been named hepatic leukemia factor (*HLF*) (Inaba et al. 1992; Hunger et al. 1992). Emerging insight into the actions of oncogenic transcription factors has led to a developmental model of hematopoietic progenitor cell transformation based on the disruption of transcriptional control (Look 1995) – an hypothesis that will be illustrated in this chapter by exploring the contributions of hybrid E2A-HLF proteins to acute pro-B cell leukemia in humans.

2 Formation of E2A-HLF Chimeric Transcripts and Proteins

The *E2A-HLF* fusion gene in pro-B cell acute lymphoblastic leukemia (ALL) arises from the translocation t(17;19)(q22;p13) (Inaba et al. 1992; Hunger et al. 1992) and affects approximately 0.5%–1.0% of childhood ALLs with this immunophenotype (Raimondi 1991). Rearrangements of these two genes affect approximately half of the cases with the t(17;19) translocation, suggesting that it, like the t(1;19) translocation (Privitera 1992), is heterogenous at the molecular level (Inaba et al. 1992; Hunger et al. 1992; Hunger et al. 1994a; Devaraj et al. 1994). Consistent features of E2A-HLF-associated leukemias include a pro-B cell immunophenotype lacking cytoplasmic or surface immunoglobulin heavy chain expression, onset of leukemia in early adolescence, hypercalcemia and disseminated intravascular coagulation at diagnosis, and a poor prognosis even with intensive multiagent ALL chemotherapy (Inaba et al. 1992; Hunger et al. 1992; 1994a; Ohyashiki et al. 1991). Although cases with *E2A-HLF* chimeric genes are quite rare, additional examples were recently identified, including two in which alternative E2A-HLF fusion proteins were detected (Hunger et al. 1994a).

Two types of genomic rearrangements leading to *E2A-HLF* fusion genes have been reported (Inaba et al. 1995; Hunger et al. 1992, 1994a). In the most common type, the breakpoints occur between *E2A* intron 13 and *HLF* intron 3. Because exon 13 of *E2A* and exon 4 of *HLF* would be joined in different translational reading frames by simple fusions of these two exons, a special joining region comprised of genomic sequences from intron 13 of *E2A* and N-region sequences, presumably added by terminal deoxynucleotidyl transferase, is included in the fusion mRNA between the coding sequences of these two genes. In the second type of *E2A-HLF* gene fusion, the breakpoints occur within *E2A* intron 12 and *HLF* intron 3, which are in the same reading frame, so that direct splicing yields an in-frame fusion transcript (Hunger et al. 1994a). The latter type of *E2A-HLF* fusion includes the same sequences of *HLF*, but not *E2A* exon 13 or the cryptic exon formed at the breakpoint of the most common type of translocations affecting these two genes. In both versions, *E2A* sequences are joined with those of the *HLF* gene, which belongs to the basic region/leucine zipper (bZIP) superfamily of transcription factors (Vinson et al. 1989) (Fig. 1).

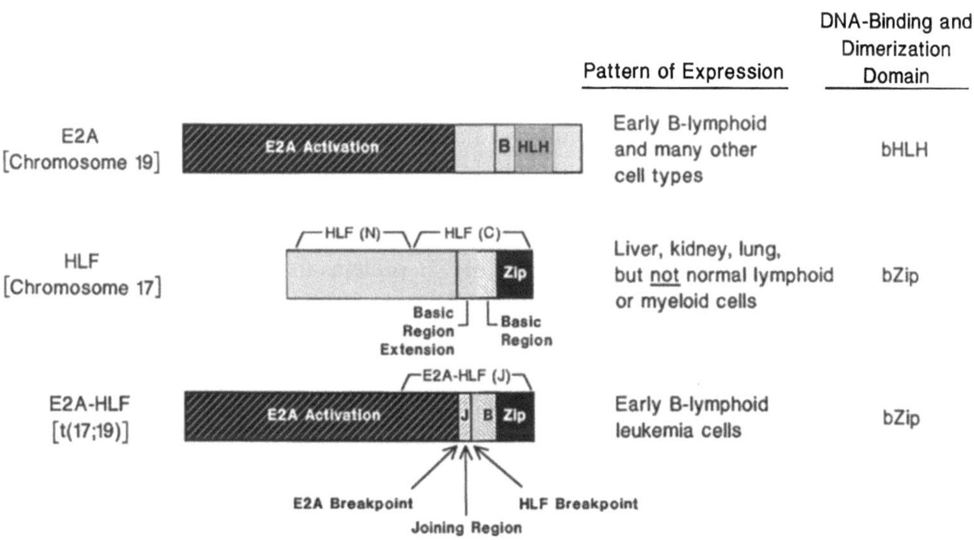

Fig. 1. The recently discovered E2A-HLF hybrid transcription factor. This protein, a result of the t(17;19) translocation in pro-B lymphoblasts, combines the *trans*-activation domain of the E2A protein with the basic-region/leucine-zipper (bZIP) DNA-binding and dimerization domain of HLF, a member of the bZIP family that normally regulates gene expression in hepatocytes, brain and renal cells. The chimera may bind to DNA sequences normally recognized by the HLF protein in the liver, brain and kidney, or perhaps competes with a close homologue for a vital developmental gene in pro-B lymphoid cells. The effector (or *trans*-activator) region appears to function in the same manner as it does in its normal context as part of the E2A protein. Thus, downstream responder genes may be dysregulated, leading to the development of ALL. (From Inaba et al. 1994)

3 E2A – A Master Regulator of B Cell Development

The *E2A* gene was first identified by MURRE et al. (1989a), who cloned one of the cDNAs it encodes by virtue of the fact that it encodes a protein (E12) that binds to the κ E2 regulatory site of the immunoglobulin κ light chain gene promoter. *E2A* was subsequently shown to encode three differentially spliced products, E12, E47 and E2-5, each of which belongs to the bHLH family of transcriptional regulatory proteins (MURRE et al. 1989a,b; SUN and BALTIMORE 1991; HENTHORN et al. 1990a,b). The bHLH domain is encoded by ca. 60 amino acids that form a structural domain consisting of two amphipathic helices separated by a loop region of variable length (thus, helix-loop-helix), responsible for homo- and heterodimerization, which is preceded by a basic region responsible for sequence-specific DNA binding (MURRE et al. 1989a,b). E2A proteins are members of a family of bHLH proteins, including those encoded by the *daughterless Drosophila* gene (CRONMILLER et al. 1988; CAUDY et al. 1988) and members of the MyoD family of myogenic proteins (WEINTRAUB 1993; BUCKINGHAM 1994). DNA-binding by E2A is mediated by either homodimers or heterodimers with other bHLH proteins, with the precise binding specificity to variations of the so-called E-box sequence motif determined by the dimerization partners of each complex (BLACKWELL and WEINTRAUB 1990). Recent structural analysis has supported the experimental observations regarding the favoured conformations of homo- and heterodimers formed by the E2A bHLH domains (ELLENBERGER et al. 1994). In addition, the NH_2-terminal sequences of E2A that are included in E2A-HLF fusion proteins (Fig. 1) have been shown to contain two discrete transcriptional activation domains (HENTHORN et al. 1990a; ARONHEIM et al. 1993; QUONG et al. 1993), called ADI and ADII, the latter of which is also referred to as a loop-helix or LH activation domain.

In most tissues, E2A heterodimerizes with tissue-specific types of bHLH proteins to regulate developmentally coordinated gene expression (MURRE et al. 1989b; LASSAR et al. 1991). The tissue-specific bHLH proteins include TAL/SCL family members, which heterodimerize with E2A (HSU et al. 1994; XIA et al. 1994) and have been shown to be essential for normal erythropoiesis during development (SHIVDASANI et al. 1995). Interestingly, the TAL/SCL proteins are capable of contributing to T cell acute leukemia when they are aberrantly expressed as a result of translocations involving the T cell receptor gene loci (BEGLEY et al. 1989; CHEN et al. 1990; XIA et al. 1991). In B cells, however, E2A is uniquely able to bind E-box sequences as a homodimeric complex (LASSAR et al. 1991; BAIN et al. 1993), apparently due to stabilization of the complex through an intermolecular disulfide bond, which is disrupted in non-B cells (BENEZRA 1994). The importance of E2A proteins in B cell development is indicated by the fact that homozygous mutant mice lacking functional E2A proteins have arrested B cell development at an early stage, but are otherwise developmentally normal (BAIN et al. 1994; ZHUANG et al. 1994).

An important contribution of the *E2A* gene to E2A-HLF fusion proteins centers around the fact that it is expressed early in B cell development, because the *E2A* promoter drives expression of the chimera in leukemic cells harboring the t(17;19) translocation. In addition, the inclusion of both E2A transcriptional activator regions within NH_2-terminal sequences of the fusion protein (Fig. 1) appears to be critical for its ability to transform cells. By contrast, the bHLH domain of E2A does not appear to influence leukemogenesis by the E2A-HLF oncoprotein, because this region is not included in the chimera, as it has been replaced by the bZIP DNA binding and dimerization domain of HLF.

4 HLF and Members of the PAR Family of bZIP Proteins

The *HLF* gene encodes a member of the basic region/leucine zipper (bZIP) superfamily of transcription factors (VINSON et al. 1989), which bind DNA as homodimers and heterodimers held together by their leucine zipper domains. HLF belongs to the proline- and acidic amino acid-rich (PAR) bZIP subfamily (DROLET et al. 1991; KHATIB et al. 1994), which also includes DBP (MUELLER et al. 1990), an albumin gene promoter D-box binding protein, and TEF (thyrotroph embryonic factor), a protein expressed concurrently with the thyroid-stimulating hormone β gene during anterior pituitary development (DROLET et al. 1991). HLF has recently been shown to encode two proteins from alternatively spliced transcripts that are regulated by different promoters (FALVEY et al. 1995). One isoform is abundant in brain, liver and kidney, while the other is restricted to hepatocytes; these proteins accumulate with different circadian patterns in the liver and have distinct promoter preferences in *trans*-activation experiments. The other known mammalian PAR bZIP proteins (TEF and DBF) have been shown to *trans*-activate reporter gene expression in each cell type tested, while HLF appears more restricted, in that it *trans*-activates gene expression in CV1 and 293 cells, but not in mouse NIH-3T3 fibroblasts, human HepG2 hepatic carcinoma or Nalm-6 early B lineage leukemia cells (INABA et al. 1994; HUNGER et al. 1994b).

HLF and E2A-HLF recombinant proteins produced in vitro bind to the consensus sequence 5′-GTTACGTAAT-3′, as do E2A-HLF proteins in nuclear extracts of a leukemic cell line harboring the t(17;19) (INABA et al. 1994; HUNGER et al. 1994b). E2A-HLF functions through this sequence motif as a potent *trans*-activator of reporter gene expression in pro-B cells. Although the UOC-B1 cell line that expresses E2A-HLF also expresses mRNA for TEF and DBP, both of the latter genes have been reported to be extensively posttranscriptionally regulated (DROLET et al. 1991; MUELLER et al. 1990), and neither TEF nor DBP, nor HLF, was detectable in leukemic cells in studies employing sensitive immunoprecipitation techniques (INABA et al. 1994). Furthermore, because the E2A-HLF chimeric proteins lacks the NH_2-terminal extension of the HLF basic region that characterizes members of the PAR bZIP subfamily, E2A-HLF homodimers are even more restricted than HLF in their binding to the primary HLF consensus recognition sequence (GTTACGTAAT) (HAAS et al. 1995). Research to identify the targets of

E2A-HLF binding has yielded provocative clues. In the Baf-3 line of interleukin-3 (IL3)-dependent murine pro-B cells, overexpression of the fusion protein prolonged cell survival after withdrawal of IL3, suggesting a primary effect on genes responsible for the prevention of lineage-specific apoptosis (INABA et al. 1995).

5 Transformation Mediated by E2A-HLF

The oncogenic potential of E2A-HLF was recently established in murine NIH-3T3 cells, in which the fusion protein induced anchorage-independent growth and rendered the cells tumorigenic in nude mice (YOSHIHARA et al. 1995). Proteins lacking the *trans*-activation domain of E2A or the leucine zipper dimerization domain of HLF were inactive, demonstrating a requirement for both elements in cell transformation. What is the role of E2A-HLF fusion proteins in pro-B cells? Do they function primarily as DNA-binding transcription factors or in protein-protein interactions? The chimeric protein appears to bind preferentially as a homodimer in leukemic cells, suggesting that its transcriptional regulatory effects may not require cross-dimerization with other bZIP or more divergent proteins (INABA et al. 1994). Taken together, these findings suggest a model in which homodimeric E2A-HLF DNA-binding complexes positively subvert transcriptional programs that normally are quiescent or actively repressed during lymphoid cell development. Whether the gene targets of E2A-HLF-mediated transformation contain binding sites recognized by HLF in liver and kidney or similar sites ordinarily bound by related transcription factors in developing B lineage cells remains a subject of future research.

6 E2A-HLF in Development

Although, for the most part, the mechanisms underlying the roles of chimeric genes in tumorigenesis remain unknown and are the focus of current research, intriguing similarities have emerged between conserved regions of mammalian transcription factors and those of "master" developmental proteins regulating the earliest stages of *Drosophila* embryogenesis (LOOK 1995; NUSSLEIN-VOLLHARD 1980,1987). These findings suggest a developmental model of progenitor cell oncogenesis based on the disruption of transcriptional control – an hypothesis that underlies our research into the contribution of hybrid E2A-HLF proteins to acute pro-B cell leukemia in humans.

References

Aronheim A, Shiran R, Rosen A, Walker MD (1993) The E2A gene product contains two separable and functionally distinct transcription activation domains. Proc Natl Acad Sci USA 90: 8063–8067

Bain G, Gruenwald S, Murre C (1993) E2A and E2-2 are subunits of B-cell-specific E2-box DNA-binding proteins. Mol Cell Biol 13: 3522–3529

Bain G, Robanus-Maandag EC, Izon DJ, Amsen D, Kruisbeek AM, Weintraub BC, Krop I, Schlissel MS, Feeney AJ, van Roon M, van der Valk M, te Riele HPJ, Berns A, Murre C (1994) E2A proteins are required for proper B-cell development and initiation of immunoglobulin gene rearrangements. Cell 79: 885–892

Begley CG, Aplan PD, Davey MP, Nakahara K, Tchorz K, Kurtzberg J, Hershfield MS, Haynes BF, Cohen DI, Waldmann TA et al. (1989) Chromosomal translocation in a human leukemic stem-cell line disrupts the T-cell antigen receptor delta-chain diversity region and results in a previously unreported fusion transcript. Proc Natl Acad Sci USA 86: 2031–2035

Benezra R (1994) An intermolecular disulfide bond stabilizes E2A homodimers and is required for DNA binding at physiological temperatures. Cell 79: 1057–1067

Blackwell TK, Weintraub H (1990) Differences and similarities in DNA-binding preferences of MyoD and E2A protein complexes revealed by binding site selection. Science 250: 1104–1110

Buckingham M (1994) Molecular biology of muscle development. Cell 78: 15–21

Caudy M, Vassin H, Brand M, Tuma R, Jan LY, Jan YN (1988) daughterless, a Drosophila gene essential for both neurogenesis and sex determination, has sequence similarities to myc and the achaete- scute complex. Cell 55: 1061–1067

Chen Q, Cheng JT, Tasi LH, Schneider N, Buchanan G, Carroll A, Crist W, Ozanne B, Siciliano MJ, Baer R (1990) The tal gene undergoes chromosome translocation in T cell leukemia and potentially encodes a helix-loop-helix protein. EMBO J 9: 415–424

Cronmiller C, Schedl P, Cline TY (1988) Molecular characterization of daughterless, a Drosophila sex determination gene with multiple roles in development. Genes Dev 2: 1666–1676

Devaraj PE, Foroni L, Sekhar M, Butler T, Wright F, Mehta A, Samson D, Prentice HG, Hoffbrand AV, Secker-Walker LM (1994) E2A/HLF fusion cDNAs and the use of RT-PCR for the detection of minimal residual disease in t(17;19) (q22;p13) acute lymphoblastic leukemia. Leukemia 8: 1131–1138

Downing JR, Look AT (1995) MLL fusion genes in the 11q23 acute leukemias. In: Freireich EJ, Kantarjian H (eds) Leukemia: advances in research and treatment. Kluwer Academic, Boston, pp 73–91

Drolet DW, Scully KM, Simmons DM, Wegner M, Chu KT, Swanson LW, Rosenfeld MG (1991) TEF, a transcription factor expressed specifically in the anterior pituitary during embryogenesis, defines a new class of leucine zipper proteins. Genes Dev 5: 1739–1753

Ellenberger T, Fass D, Arnaud M, Harrison SC (1994) Crystal structure of transcription factor E47: E-box recognition by a basic region helix-loop-helix dimer. Genes Dev 8: 970–980

Falvey E, Fleury-Olela F, Schibler U (1995) The rat hepatic leukemia factor (HLF) gene encodes two transcriptional activators with distinct circadian rhythms, tissue distributions and target preferences. EMBO J 14: 4307–4317

Haas NB, Cantwell CA, Johnson PF, Burch JBE (1995) DNA-binding specificity of the PAR basic leucine zipper protein VBP partially overlaps those of the C/EBP and CREB/ATF families and is influenced by domains that flank the core basic region. Mol Cell Biol 15: 1923–1932

Henthorn P, Kiledjian M, Kadesch T (1990a) Two distinct transcription factors that bind the immunoglobulin enhancer E5/E2 motif. Science 247: 467–470

Henthorn P, McCarrick-Walmsley R, Kadesch T (1990b) Sequence of the cDNA encoding ITF-1, a positive-acting transcription factor. Nucleic Acids Res 18: 677

Hunger SP, Ohyashiki K, Toyama K, Cleary ML (1992) HLF, a novel hepatic bZIP protein, shows altered DNA-binding properties following fusion to E2A in t(17;19) acute lymphoblastic leukemia. Genes Dev 6: 1608–1620

Hunger SP, Devaraj PE, Foroni L, Secker-Walker LM, Cleary ML, (1994a) Two types of genomic rearrangements create alternative E2A-HLF fusion proteins in t(17;19)-ALL. Blood 83: 2261–2267

Hunger SP, Brown R, Cleary ML (1994b) DNA-binding and transcriptional regulatory properties of hepatic leukemia factor (HLF) and the t(17;19) acute lymphoblastic leukemia chimera E2A-HLF. Mol Cell Biol 14: 5986–5996

Hsu HL, Huang L, Tsan JT, Funk W, Wright WE, Hu JS, Kingston RE, Baer R (1994) Preferred sequences for DNA recognition by the TAL1 helix-loop-helix proteins. Mol Cell Biol 14: 1256–1265

Inaba T, Shapiro LH, Funabiki T, Sinclair AE, Jones BG, Ashmun RA, Look AT (1994) DNA-binding specificity and trans-activating potential of the leukemia-associated E2A-Hepatic Leukemia Factor fusion protein. Mol Cell Biol 14: 3403–3413

Inaba T, Roberts WM, Shapiro LH, Jolly KW, Raimondi SC, Smith SD, Look AT (1992) Fusion of the leucine zipper gene HLF to the E2A gene in human acute B-lineage leukemia. Science 257: 531–534

Inaba T, Inukai T, Yoshihara T, Seyschab H, Ashmun RA, Canman CE, Laken SJ, Kastan MB, Look AT (1996) Reversal of apoptosis by the leukaemia-associated E2A-HLF chimaeric transcription factor. nature (in press)

Izraeli S, Kovar H, Gadner H, Lion T (1992) Unexpected heterogeneity in E2A/PBX1 fusion messenger RNA detected by the polymerase chain reaction in pediatric patients with acute lymphoblastic leukemia. Blood 80: 1413–1417

Kamps MP, Murre C, Sun XH, Baltimore D (1990) A new homeobox gene contributes the DNA binding domain of the t(1;19) translocation protein in pre-B ALL. Cell 60: 547–555

Khatib ZA, Inaba T, Valentine M, Look AT (1994) Chromosomal localization and cDNA cloning of the human DBP and TEF genes. Genomics 23: 344–351

Lassar AB, Davis RL, Wright WE, Kadesch T, Murre C, Voronova A, Baltimore D, Weintraub H (1991) Functional activity of myogenic HLH proteins requires hetero-oligomerization with E12/E47-like proteins in vivo. Cell 66: 305–315

LeBrun DL, Cleary ML (1994) Fusion with E2A alters the transcriptional properties of the homeodomain protein PBX1 in t(1;19) leukemias. Oncogene 9: 1641–1647

Look AT (1995) Oncogenic role "master" transcription factors in human leukemias and sarcomas: a developmental model. In: Vande Woude G (ed) Advances in cancer research. Academic, San Diego, pp 25–55

Lu Q, Wright DD, Kamps MP (1994) Fusion with E2A converts the Pbx1 homeodomain protein into a constitutive transcriptional activator in human leukemias carrying the t(1:19) translocation. Mol Cell Biol 14: 3938–3948

Mueller CR, Maire P, Schibler U (1990) DBP, a liver-enriched transcriptional activator, is expressed late in ontogeny and its tissue specificity is determined posttranscriptionally. Cell 61: 279–291

Murre C, McCaw PS, Baltimore D (1989a) A new DNA binding and dimerization motif in immunoglobulin enhancer binding, daughterless, MyoD, and myc proteins. Cell 56: 777-783

Murre C, McCaw PS, Vaessin H, Caudy M, Jan LY, Cabrera CV, Buskin JN, Hauschka SD, Lassar AB, Weintraub H, Baltimore D (1989b) Interactions between heterologous helix-loop-helix proteins generate complexes that bind specifically to a common DNA sequence. Cell 58: 537–544

Nourse J, Mellentin JD, Galili N, Wilkinson J, Stanbridge E, Smith SD, Cleary ML (1990) Chromosomal translocation t(1;19) results in synthesis of a homeobox fusion mRNA that codes for a potential chimeric transcription factor. Cell 60: 535–545

Numata S, Kato K, Horibe K (1993) New E2A/PBX1 fusion transcript in a patient with t(1;19) (q23; p13) acute lymphoblastic leukemia. Leukemia 7: 1441

Nusslein-Volhard C, Wieschaus E (1980) Mutations affecting segment number and polarity in Drosophila. Nature 287: 795–801

Nusslein-Volhard C, Frohnhofer HG, Lehmann R (1987) Determination of anteroposterior polarity in Drosophila. Science 238: 1675–1681

Ohyashiki K, Fujieda H, Miyauchi J, Ohyashiki JH, Tauchi T, Saito M, Nakazawa S, Abe K, Yamamoto K, Clark SC et al. (1991) Establishment of a novel heterotransplantable acute lymphoblastic leukemia cell line with a t(17;19) chromosomal translocation the growth of which is inhibited by interleukin-3. Leukemia 5: 322–331

Privitera E, Kamps MP, Hayashi Y, Inaba T, Shapiro LH, Raimondi SC, Behm F, Hendershot L, Carroll AJ, Baltimore D, Look AT (1992) Different molecular consequences of the 1;19 chromosomal translocation in childhood B-cell precursor acute lymphoblastic leukemia. Blood 79: 1781–1788

Quong MW, Massari ME, Zwart R, Murre C (1993) A new transcriptional-activation motif restricted to a class of helix-loop-helix proteins is functionally conserved in both yeast and mammalian cells. Mol Cell Biol 13: 792–800

Rabbitts TH (1994) Chromosomal translocations in human cancer. Nature 372: 143–149

Raimondi SC, Privitera E, Williams DL, Looks AT, Behm F, Rivera GK, Crist WM, Pui C (1991) New recurring chromosomal translocations in childhood acute lymphoblastic leukemia. Blood 77: 2016–2022

Rowley JD, Aster JC, Sklar J (1993) The clinical applications of new DNA diagnostic technology on the management of cancer patients. J Am Med Assoc 270: 2331–2337

Shivdasani RA, Mayer EL, Orkin SH (1995) Absence of blood formation in mice lacking the T-cell leukaemia oncoprotein tal-1/SCL. Nature 373: 432–434

Sun XH, Baltimore D (1991) An inhibitory domain of E12 transcription factor prevents DNA binding in E12 homodimers but not in E12 heterodimers. Cell 64: 459–470

Van Dijk MA, Voorhoeve PM, Murre C (1993) Pbx1 is converted into a transcriptional activator upon acquiring the N-terminal region of E2A in pre-B-cell acute lymphoblastoid leukemia. Proc Natl Acad Sci USA 90: 6061–6065

Vinson CR, Sigler PB, McKnight SL (1989) Scissors-grip model for DNA recognition by a family of leucine zipper proteins. Science 246: 911–916

Weintraub H (1993) The MyoD family and myogenesis: redundancy, networks, and thresholds. Cell 75: 1241–1244

Xia Y, Brown L, Yang CY, Tsan JT, Siciliano MJ, Espinosa III R, Le Beau MM, Baer RJ (1991) TAL2, a helix-loop-helix gene activated by the (7;9) (q34; q32) translocation in human T-cell leukemia. Proc Natl Acad Sci USA 88: 11416–11420

Xia Y, Hwang LH, Cobb MH, Baer RJ (1994) Products of the TAL2 oncogene in leukemic T cells: bHLH phosphoproteins with DNA-binding activity. Oncogene 9: 1437–1446

Yoshihara T, Inaba T, Shapiro LH, Kato J, Look AT (1995) E2A-HLF-mediated cell transformation requires both the trans-activation domain of E2A and the leucine zipper dimerization domain of HLF. Mol Cell Biol 15: 3247–3255

Zhuang Y, Soriano P, Weintraub H (1994) The helix-loop-helix gene E2A is required for B-cell formation. Cell 79: 875–884

Transcription Factors of the bHLH and LIM Families: Synergistic Mediators of T Cell Acute Leukemia?

R. BAER[1], L.-Y. HWANG[1], and R.O. BASH[2]

1 Introduction

Patients with T cell acute lymphoblastic leukemia (T-ALL) often harbor tumor-specific chromosome translocations in their malignant cells (reviewed by RAIMONDI 1993). In an effort to understand the etiology of T-ALL, many investigators have sought to identify the genes that are altered as a consequence of these chromosomal defects (RABBITTS 1994). To date these studies have uncovered nine presumptive proto-oncogenes, each of which can be activated in T-ALL cells by aberrant juxtaposition with the T cell receptor sequences on chromosomes 7 or 14 (HWANG and BAER 1995). For instance, the (8;14) (q24;q11) translocation serves to deregulate the *MYC* gene on chromosome 8 by recombining it with the T cell receptor α/δ chain locus on chromosome 14. The various proto-oncogenes implicated in T-ALL are listed in Table 1, along with the major chromosome translocations that are responsible for their activation.

[1]Molecular Immunology Center, Department of Microbiology, UT Southwestern Medical Center at Dallas, 6000 Harry Hines Blvd., Dallas, TX 75235-9140, USA
[2]Molecular Immunology Center, Department of Pediatrics, UT Southwestern Medical Center at Dallas, 6000 Harry Hines Blvd., Dallas, TX 75235-9140, USA

Table 1. The proto-oncogenes activated by chromosomal rearrangement in T cell acute lymphoblastic leukemia

Chromosome Translocation	Frequency in T-ALL(%)	Gene	Amino Acid Motif	Probable function
t(8;14)(q24;q11)	2	*MYC*	bHLH-ZIP	Transcription factor
t(10;14)(q24;q11)	4	*HOX11*	Homeodomain	Transcription factor
t(11;14)(p15;q11)	1	*LMO1*	LIM	Transcription factor
t(11;14)(p13;q11)	7	*LMO2*	LIM	Transcription factor
t(1;14)(p34;q11)	3	*TAL1*	bHLH	Transcription factor
t(7;9)(q34;q32)	2	*TAL2*	bHLH	Transcription factor
t(7;19) (q34;p13)	<1	*LYL1*	bHLH	Transcription factor
t(7;9) (q34;q34)	2	*TAN1*	notch	Signal transduction
t(1;7)(p32;q34)	1	*LCK*	PTK	Signal transduction

The major chromosome translocation responsible for the activation of each proto-oncogene is listed, along with an estimate of its frequency among T-ALL patients. The proto-oncogenes are described according to gene symbols accepted by the HUGO Nomenclature Committee; alternative designations have been used for *HOX11 (TCL3)*, *LMO1 (RBTN1/TTG1)*, *LMO2 (RBTN2/TTG2)*, and *TAL1 (TCL5, tal, SCL)*. With the exception of TAN1 and LCK, the protein products of these genes possess amino acid motifs characteristic of known transcription factors (see text). *TAN1* encodes a homolog of the *Drosophila* notch protein, and *LCK* encodes a protein tyrosine kinase (PTK).
T-ALL, T cell acute lymphoblastic leukemia

Some hematopoietic malignancies are characterized by common chromosomal defects that serve to activate a particular proto-oncogene in a vast majority (>95%) of the affected patients; prominent examples include the translocations that deregulate *MYC* in Burkitt's lymphoma or *BCL2* in follicular B cell lymphoma (Gaidano and Dalla-Favera 1993; Gauwerky and Croce 1993). In contrast, each of the chromosome translocations listed in Table 1 is found in a small minority (<10%) of T-ALL patients. The absence of a more common genetic lesion seems at odds with the homogeneity of presentation features and clinical outcomes shared by most T-ALL patients (Pui 1995). Nevertheless, recent studies imply the existence of a major pathway of T-ALL development which involves the ectopic activation of certain lineage-specific transcription factors.

As indicated in the table, most of the genes implicated in T-ALL encode proteins with amino acid motifs characteristic of known transcription factors. These include the combined basic helix-loop-helix leucine zipper (bHLH-Zip) motif of MYC, the homeodomain (found in HOX11), the LIM domain (in LMO1 and LMO2), and the basic helix-loop-helix (bHLH) domain (in TAL1, TAL2, and LYL1) (Table 1). This review will focus on the LIM and bHLH proteins; recent work suggests that these transcription factors play complementary roles in normal development and promote a common pathway of leukemogenesis in patients with T-ALL.

2 The LIM Proteins Implicated in T Cell Acute Lymphoblastic Leukemia

The LIM motif is a cysteine-rich sequence of approximately 60 amino acids that was originally found in a subset of homeodomain transcription factors (Way and Chalfie 1988; Freyd et al. 1990; Karlsson et al. 1990). More than a dozen "LIM-homeodomain" (LIM-HD) proteins have already been described, many of which control different aspects of cell-type determination during early embryogenesis (reviewed by SANCHEZ-GARCIA and RABBITTS 1994). The DNA-binding activity of LIM-HD proteins is mediated by the homeodomain, and their *trans*-activation potential generally resides in sequences outside the LIM moiety. Hence, the functional role of the LIM motif has been the subject of speculation.

The LIM motif consists of two tandem metal-binding modules, each of which employs four conserved amino acid residues (primarily cysteines) to coordinate a single Zn(II) ion (SADLER et al. 1992; MICHELSEN et al. 1993; ARCHER et al. 1994; KOSA et al. 1994). The backbone conformation of the COOH-terminal module resembles the DNA-binding domains of the glucocorticoid receptor and GATA-1 transcription factors (PEREZ-ALVARADO et al. 1994). Despite this structural similarity, direct evidence of DNA recognition by LIM sequences has not been reported. Instead, LIM domains have been shown to mediate diverse types of protein-protein interaction, including both homotypic and heterotypic LIM/LIM assembly (FEUERSTEIN et al. 1994; SCHMEICHEL and BECKERLE 1994), as well as specific association with distinct amino acid motifs (see below).

The *RBTN1/TTG1* and *RBTN2/TTG2* genes were identified by analysis of chromosome translocations associated with T-ALL (Table 1). These genes encode small polypeptides (156 and 158 amino acids, respectively) that are comprised almost entirely of two tandem LIM motifs (BOEHM et al. 1988; MCGUIRE et al. 1989; BOEHM et al. 1991; ROYER-POKORA et al. 1991). As such, they represent founding members of a distinct class of LIM proteins (the "LIM-only" proteins) that lack an associated homeodomain or, indeed, any other recognizable amino acid motif (SANCHEZ-GARCIA and RABBITTS 1993). *RBTN1/TTG1* and *RBTN2/TTG2* were recently assigned new gene symbols, *LMO1* and *LMO2* (for LIM-only proteins 1 and 2), by the HUGO Nomenclature Committee (P. McAlpine, personal communication). The proteins encoded by *LMO1* and *LMO2* localize to the nucleus (MCGUIRE et al. 1991; WARREN et al. 1994), a property consistent with their proposed function as regulators of RNA transcription (BOEHM et al. 1990). Although they display distinct expression patterns during normal development, both *LMO1* and *LMO2* are either poorly transcribed or entirely quiescent in normal T lymphocytes. Hence, the chromosome translocations involving these genes (Table 1) are likely to promote leukemogenesis by inducing ectopic expression of LMO1 or LMO2 polypeptides in T lineage cells. The leukemic potential of these proteins has been fully established in transgenic models of T-ALL; hence, thymic malignancies are induced in mice by targeted T cell expression of either an *LMO1* or *LMO2* transgene (FISCH et al. 1992; MCGUIRE et al. 1992).

3 The Basic Helix-Loop-Helix Proteins

The bHLH motif is a conserved domain (55–60 amino acids in length) that mediates sequence-specific DNA recognition for a large family of transcription factors (MURRE et al. 1989; KADESCH 1993). Individual bHLH polypeptides do not bind

DNA by themselves; instead, they associate with other bHLH polypeptides to form homodimers or heterodimers with DNA-binding activity. The bHLH domain has the potential to form two amphipathic α helices separated by an intervening loop (MURRE et al. 1989). Upon protein dimerization the associated bHLH domains fold into a parallel four-helix bundle that is stabilized by a hydrophobic core of conserved amino acid sidechains (FERRE-D'AMARE et al. 1993).

The various bHLH transcription factors can be classified into different groups on the basis of their functional properties, expression patterns, and degrees of amino acid sequence homology. The earliest classification scheme described three groups of bHLH proteins (MURRE et al. 1989). Class A proteins, which are expressed in a broad spectrum of tissues and cell types, have the ability to self-associate into homodimers with DNA-binding and *trans*-activation potential. Four distinct class A bHLH proteins are found in mammals: E47, E12, E2–2, and HEB. In contrast, class B proteins, which exhibit tissue-specific patterns of expression, do not homodimerize effectively. Hence, the functional properties of class B polypeptides are dependent on class A proteins, with which they readily form bHLH heterodimers. Class B proteins include muscle-specific transcription factors that drive mammalian myogenesis (e.g., MyoD1), the proneural proteins encoded by the *Drosophila achaete-scute* locus, and the TAL1-related proteins implicated in T-ALL (see below). Class C proteins (such as MYC) are distinct in that they possess tandem bHLH and leucine zipper motifs. In general, the bHLH-Zip proteins do not interact with class A or class B polypeptides, but instead form homo- or heterodimeric complexes with other class C proteins. Additional groups of bHLH proteins are now recognized, and these have been reviewed comprehensively by MURRE et al. (1994).

4 TAL1, TAL2, and LYL1: The Basic-Helix-Loop-Helix Proteins Implicated in T Cell Acute Lymphoblastic Leukemia

Three genes encoding class B bHLH proteins were identified by analysis of the chromosome translocations associated with T-ALL (Table 1) (HWANG and BAER 1995). During normal development, these genes (*TAL1, TAL2,* and *LYL1*) display different patterns of tissue-specific expression; each, however, appears to be transcriptionally inactive within the major populations of thymocytes and peripheral T cells (KUO et al. 1991; XIA et al. 1991; MOUTHON et al. 1993; KALLIANPUR et al. 1994; PULFORD et al. 1995). Thus, in a manner reminiscent of the *LMO* genes, the chromosome abnormalities involving *TAL1, TAL2* and *LYL1* are likely to promote leukemogenesis by inducing ectopic expression of their gene products in T lineage cells. Moreover, the bHLH domains encoded by these genes exhibit a remarkable level of sequence homology (>85% amino acid identity). Hence, activation of either *TAL1, TAL2* or *LYL1* probably constitutes an equivalent step in the pathogenesis of T-ALL (BAER 1993).

Chromosome translocations involving the *TAL1*-related genes are found in fewer than 5% of T-ALL patients. Nevertheless, an additional 20%–25% of these patients carry tumor-specific alterations of *TAL1* (termed tall^{d} rearrangements) that are not detected cytogenetically (BROWN et al. 1990; BERNARD et al. 1991; APLAN et al. 1992; BASH et al. 1993). The tall^{d} rearrangements arise from site-specific deletions that remove approximately 90 kilobasepairs of upstream sequence from the *TAL1* gene, including transcriptional regulatory elements that are normally repressed in T cells (BROWN et al. 1990; APLAN et al. 1990). As a result, the coding exons of *TAL1* are juxtaposed with the 5′-untranslated exon of *SIL*, a neighboring gene which, unlike *TAL1*, is transcriptionally active in T lymphocytes (APLAN et al. 1991). The tall^{d} allele encodes chimeric *SIL/TAL1* transcripts that initiate from the SIL promoter but direct the synthesis of TAL1 polypeptides. Hence, the tall^{d} rearrangement is functionally comparable to the chromosome translocations involving *TAL1* in that it induces ectopic expression of TAL1 polypeptides in T lineage cells.

Malignant expression of *TAL1* is induced in approximately 25% of T-ALL patients by the aforementioned chromosome translocations and local DNA rearrangements. In addition, we recently showed that *TAL1* is ectopically expressed in the leukemic cells of most T-ALL patients (60%–65%), including many that do not display an obvious alteration of the *TAL1* gene (BASH et al. 1995). Although we have not ascertained how ectopic *TAL1* expression is induced in the absence of gross structural DNA alterations, this result suggests that *TAL1* activation represents a major pathway in the development of T-ALL. In contrast, evidence supporting a broader role for the *TAL1*-related genes (*TAL2* and *LYL1*) has yet to emerge. Nevertheless, activation of either *TAL1, TAL2* or *LYL1* occurs in a majority (at least 60%) of T-ALL patients, and as such it represents the most common oncogenic lesion associated with this disease.

5 The Properties of *TAL1* Polypeptides

At least two protein products of the *TAL1* gene are detected in the leukemic cells of T-ALL patients: a full-length polypeptide (pp42^{TAL1}; amino acid residues 1–331) and a truncated species (pp22^{TAL}; residues 176–331) (CHENG et al. 1993). Both polypeptides are phosphorylated in vivo, and both contain the bHLH motif. The dimerization and DNA-binding properties of the TAL1 polypeptides resemble those of other class B bHLH proteins. Thus, TAL1 polypeptides do not homodimerize effectively; instead, they preferentially associate with class A bHLH proteins to form heterodimers (e.g., TAL1/E47) that bind DNA in a sequence-specific manner and activate transcription of cognate reporter genes (HSU et al. 1991,1994a,c). These heterodimers are found in normal hematopoietic cells undergoing erythroid differentiation (Condorelli et al. 1995) as well as in the leukemic cells of T-ALL patients (Hsu et al. 1994b). Therefore, the normal and malignant properties of TAL1 are both presumably mediated by its interaction with class A bHLH proteins such as E47.

6 Collaboration Between *TAL1* and *LMO2*: Genetic Evidence

At the cellular level, tumor development usually involves the accumulation of multiple genetic lesions which, together, elicit the full malignant phenotype (Hunter 1991). In this regard, it is noteworthy that a subset of T-ALL patients harbor tumor-specific alterations of both the *TAL1* and *LMO2* genes (Wadman et al. 1994). Typically, the malignant cells of these patients carry an (11;14) (p13;q11) translocation involving *LMO2*, along with a tal^d rearrangement of the *TAL1* locus. The recurrent coactivation of *TAL1* and *LMO2* in unrelated patients strongly suggests that these genes synergistically promote the pathogenesis of T-ALL. Furthermore, gene targeting experiments also support a cooperative relationship between *TAL1* and *LMO2* during normal hematopoietic development. Hence, mice deficient for either gene display a severe failure in erythropoiesis that incurs embryonic lethality (Warren et al. 1994; Shivdasani et al. 1995). Moreover, both genes are coexpressed during normal erythroid development, and their protein products co-localize to the nuclei of erythroid progenitors (Warren et al. 1994).

7 Collaboration Between *TAL1* and *LMO2*: Physical Evidence

Coimmunoprecipitation experiments and two-hybrid assays have shown that the TAL1 and LMO2 polypeptides avidly associate with one another in vivo (Valge-Archer et al. 1994; Wadman et al. 1994). Indeed, stable complexes containing both proteins are readily detected in erythroid cells and in the leukemic cells of T-ALL patients. Thus, the genetic relationship between *TAL1* and *LMO2* during both normal (i.e., erythroid) and malignant (T-ALL) development presumably reflects a functional interaction between their protein products.

The TAL1/LMO2 interaction requires sequences within the bHLH and LIM motifs of the respective polypeptides (Wadman et al. 1994). Thus, the bHLH domain of TAL1 has the potential to associate with the LIM motifs of LMO2 as well as bHLH sequences of class A proteins such as E47. However, these interactions are not mutually exclusive; two-hybrid experiments indicate that TAL1 can interact simultaneously with both LMO2 and E47 to form a ternary complex (LMO2/TAL1/E47) (Wadman et al. 1994). The ability of LMO2 to associate with assembled TAL1/E47 heterodimers implies that the transcriptional regulatory properties of TAL1 may be subject to modulation by LMO2.

OSADA et al. (1995) recently reported that LMO2 polypeptides can interact in vivo with GATA-1, a tissue-specific transcription factor that is also essential for normal erythropoiesis (PEVNY et al. 1991). GATA-1 binds DNA in a sequence-specific fashion, and it recognizes *cis*-acting regulatory elements in the promoters of most genes expressed in erythroid cells (ORKIN 1992). Significantly, LMO2 has the potential to associate simultaneously with both GATA-1 and TAL1 (OSADA et al. 1995). Therefore, LMO2 may serve to stabilize a multicomponent protein complex that includes the GATA-1 and TAL1 polypeptides as well as a bHLH dimerization partner of TAL1 (such as E47). Through the combined specificities of its distinct DNA-binding components (e.g., GATA-1 and the TAL1/E47 heterodimer) this complex may recognize and regulate a unique subset of genes that are essential for normal erythropoiesis.

8 Basic-Helix-Loop-Helix/LIM Interactions in Normal Development

Given that the interaction between TAL1 and LMO2 is mediated by sequences within their respective bHLH and LIM domains, it is important to ascertain whether other proteins bearing these motifs also interact in an analogous fashion. Initial assessments indicate that bHLH/LIM interactions are highly specific (WADMAN et al. 1994). Thus, TAL1 has the potential to associate with either of the two LIM proteins implicated in T-ALL (i.e., LMO1 and LMO2) but not with any of the other LIM proteins tested (CRP, CRIP, zyxin, testin). Likewise, LMO1 and LMO2 can associate with each of the TAL1-related polypeptides (TAL1, TAL2, and LYL1) but not with other bHLH proteins, including E12, E47, bHLH-EC2, NHLH1, MyoD1, MAX, MYC, and AP4. The striking specificity of these interactions suggests that bHLH/LIM protein complexes play functional roles during normal development. Since the *TAL1* and *LMO2* genes are both required for successful erythrogenesis, the TAL1/LMO2 complexes found in erythroid progenitors are likely to serve a critical function in this hematopoietic lineage. The other potential bHLH/LIM complexes (e.g., TAL1/LMO1 or TAL2/LMO1) may likewise contribute to normal development in distinct cellular lineages.

9 A Common Pathway in the Development of T Cell Acute Lymphoblastic Leukemia?

In addition to its normal function in erythrogenesis, the TAL1/LMO2 complex also appears to have an oncogenic potential that can be unleashed by its inappropriate appearance in T lymphocytes. This notion is supported by the fact that some

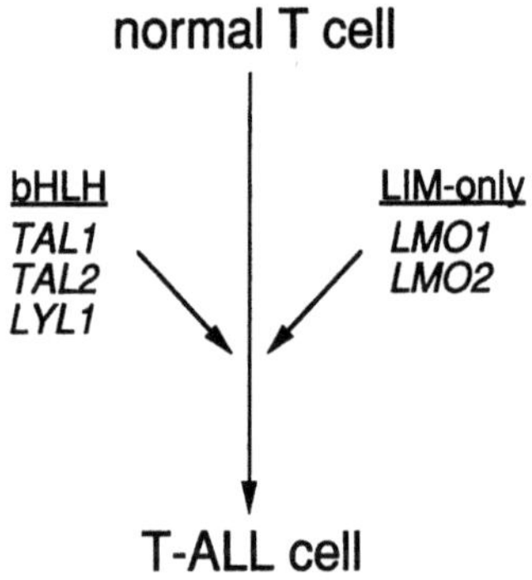

Fig. 1. A common pathway of T cell acute lymphoblastic leukemia pathogenesis? The postulated path by which a normal T lineage cell is transformed into a malignant T-ALL cell. Transformation along this pathway can be driven either independently or synergistically by the ectopic expression of a TAL1-related bHLH protein (TAL1, TAL2, or LYL1) and a leukemic LIM-only protein (LMO1 or LMO2). Additional genetic lesions presumably contribute to T-ALL development along this pathway

T-ALL patients possess tumor specific rearrangements of both *TAL1* and *LMO2* (Wadman et al. 1994), and it should prove to be testable experimentally by targeted coexpression of *TAL1* and *LMO2* transgenes in mice (Larson et al. 1995). However, many T-ALL patients display oncogenic rearrangements at only one of these two loci, implying that either gene can independently promote a common pathway of T cell leukemogenesis (Fig. 1). In this scheme, synergism between activated alleles of *TAL1* and *LMO2* may facilitate, but would not be required for, the development of T-ALL.

Finally, given that both LMO1 and LMO2 can potentially interact with any of the TAL1-related proteins, it is reasonable to ask whether other combinations of LIM and bHLH proteins can also promote T cell leukemogenesis in a cooperative fashion. In this regard, it is intriguing that the T-ALL-derived cell line RPMI-8402 has an (11;14)(p15;q11) translocation that activates *LMO1* (Boehm et al. 1988; McGuire et al. 1989) as well as a tal1^{d} rearrangement of the *TAL1* locus (Brown et al. 1990). Thus, any combination of leukemic LIM protein (i.e., LMO1 or LMO2) and TAL1-related bHLH protein (TAL1, TAL2 or LYL1) may have the potential to promote T-ALL synergistically.

Acknowledgements. We are grateful to Susan Campbell for preparing the manuscript. This work was supported by a grant from the National Cancer Institute (RO1 CA46593). R. Baer is a member of the Simmons Arthritis Research Center and a recipient of a Faculty Research Award from the American Cancer Society (FRA-421). L-Y. Hwang is supported in part by a fellowship from the Leukemia Association of North Central Texas, and R. Bash is supported by an award from the National Cancer Institute (K08 CA63544).

References

Aplan PD, Lombardi DP, Ginsberg AM, Cossman J, Bertness VL and Kirsch IR (1990) Disruption of the human SCL locus by "illegitimate" V-(D)-J recombinase activity. Science 250: 1426–1429

Aplan PD, Lombardi DP and Kirsch IR (1991) Structural characterization of *SIL*, a gene frequently disrupted in T-cell acute lymphoblastic leukemia. Mol Cell Biol 11: 5462–5469

Aplan PD, Lombardi DP, Reaman GH, Sather HN, Hammond GD and Kirsch IR (1992a) Involvement of the putative hematopoietic transcription factor *SCL* in T-cell acute lymphoblastic leukemia. Blood 79: 1327–1333

Archer V, Breton J, Sanchez-Garcia I, Osada H, Forster A, Thomson AJ and Rabbitts TH (1994) The cysteine-rich domains of LIM proteins RBTN and Isl-1 contain zinc but not iron. Proc Natl Acad Sci USA 91: 316–320

Baer R (1993) TAL1, TAL2, and LYL1: A family of basic helix-loop-helix proteins implicated in T cell acute leukemia. Sem Cancer Biol 4: 341–347

Bash RO, Crist WM, Shuster JJ, Link MP, Amylon M, Pullen J, Carroll AJ, Buchanan GR, Smith RG and Baer R (1993) Clinical features and outcome of T-cell acute lymphoblastic leukemia in childhood with respect to alterations at the *TAL1* locus: A Pediatric Oncology Group Study. Blood 81: 2110–2117

Bash RO, Hall S, Timmons CF, Crist WM, Amylon M, Smith RG and Baer R (1995) Does activation of the TAL1 gene occur in a majority of patients with T-cell acute lymphoblastic leukemia? A pediatric Oncology Group Study. Blood 86: 666–676

Bernard O, Lecointe N, Jonveaux P, Souyri M, Mauchauffe M, Berger R, Larsen CJ and Mathieu-Mahul D (1991) Two site-specific deletions and t(1;14) translocation restricted to human T-cell acute leukemias disrupt the 5′ part of the tal-1 gene. Oncogene 6: 1477–1488

Boehm T, Baer R, Lavenir I, Forster A, Waters JJ, Nacheva E and Rabbitts TH (1988) The mechanism of chromosomal translocation t(11;14) involving the T-cell receptor cδ locus on human chromosome 14q11 and a transcribed region of chromosome 11p15. EMBO J 7: 385–394

Boehm T, Foroni L, Kaneko Y, Perutz MF and Rabbitts TH (1991) The rhombotin family of cysteine-rich LIM-domain oncogenes: Distinct members are involved in T-cell translocations to human chromosomes 11p15 and 11p13. Proc Natl Acad Sci USA 88: 4367–4371

Boehm T, Foroni L, Kennedy M and Rabbitts TH (1990) The rhombotin gene belongs to a class of transcriptional regulators with a potential novel protein dimerization motif. Oncogene 5: 1103–1105

Brown L, Cheng J-T, Chen Q, Siciliano MJ, Crist W, Buchanan G and Baer R (1990) Site-specific recombination of the *tal*-1 gene is a common occurrence in human T cell leukemia. EMBO J 9: 3343–3351

Cheng J-T, Hsu H-L, Hwang L-Y and Baer R (1993b) Products of the *TAL1* oncogene: Basic helix-loop-helix proteins phosphorylated at serine residues. Oncogene 8: 677–683

Condorelli G, Vitelli L, Valtieri M, Marta I, Montesoro E, Lulli V, Baer R and Peschle C (1995) Coordinate expression and developmental role of Id2 protein and TAL1/E2A heterodimer in erythroid progenitor differentiation. Blood 86: 164–175

Ferre-DAmare AR, Prendergast GC, Ziff EB and Burlet SK (1993) Recognition by Max of its cognate DNA through a dimeric b/HLH/Z domain. Nature 363: 38–45

Feuerstein R, Wang X, Song D, Cooke NE and Liebhaber SA (1994) The LIM/double zinc-finger motif functions as a protein dimerization domain. Proc Natl Acad Sci USA 91: 10655–10659

Fisch P, Boehm T, Lavenir I, Larson T, Arno J, Forster A and Rabbitts TH (1992) T-cell acute lymphoblastic lymphoma induced in transgenic mice by the RBTN1 and RBTN2 LIM-domain genes. Oncogene 7: 2389–2397

Freyd G, Kim SK and Horvitz HR (1990) Novel cysteine-rich motif and homeodomain in the product of the *Caenorhabditis elegans* cell lineage gene *lin-II*. Nature 344: 876–879

Gaidano G and Dalla-Favera R (1993) Biologic and molecular characterization of non-Hodgkin's Lymphoma. Cur Opin Oncol 5: 776–784

Gauwerky CE and Croce CM (1993) Chromosomal translocations in leukaemia. Sem Cancer Biol 4: 333–340

Hsu H-L, Cheng J-T, Chen Q and Baer R (1991) Enhancer-binding activity of the tal-1 oncoprotein in association with the E47/E12 helix-loop-helix proteins. Mol Cell Biol 11: 3037–3042

Hsu H-L, Huang L, Tsan JT, Funk W, Wright WE, Hu J-S, Kingston RE and Baer R (1994a) Preferred sequences for DNA recognition by the TAL1 helix-loop-helix proteins. Mol Cell Biol 14: 1256–1265

Hsu H-L, Wadman I and Baer R (1994b) Formation of in vivo complexes between the TAL1 and E2A polypeptides of leukemic T cells. Proc Natl Acad Sci USA 91: 3181–3185

Hsu H-L, Wadman I, Tsan JT and Baer R (1994c) Positive and negative transcriptional control by the TAL1 helix-loop-helix protein. Proc Natl Acad Sci USA 91: 5947–5951

Hunter T (1991) Cooperation between oncogenes. Cell 64: 249–270

Hwang L-Y and Baer R (1995) The role of chromosome translocations in T cell acute leukemia. Cur Opin Immunol 7: 659–664

Kadesch T (1993) Consequences of heteromeric interactions among helix-loop-helix proteins. Cell Growth Differ 4: 49–55

Kallianpur AR, Jordan JE and Brandt SJ (1994) The *SCL/TAL-1* gene is expressed in progenitors of both the hematopoietic and vascular systems during embryogenesis. Blood 83: 1200–1208

Karlsson O, Thor S, Norberg T, Ohlsson H and Edlund T (1990) Insulin gene enhancer binding protein Isl-1 is a member of a novel class of proteins containing both a homeo- and a Cys-His domain. Nature 344: 879–882

Kosa JL, Michelsen JW, Louis HA, Olsen JI, Davis DR, Beckerle MC and Winge DR (1994) Common metal ion coordination in LIM domain proteins. Biochemistry 33: 468–477

Kuo SS, Mellentin JD, Copeland NG, Gilbert DJ, Jenkins NA and Cleary ML (1991) Structure, chromosome mapping, and expression of the mouse *Lyl–1* gene. Oncogene 6: 961–968

Larson RC, Osada H, Larson TA, Lavenir I and Rabbitts TH (1995) The oncogenic LIM protein Rbtn2 causes thymic developmental aberrations that precede malignancy in transgenic mice. Oncogene 11: 853–862

McGuire EA, Davis AR and Korsmeyer SJ (1991) T-cell translocation gene 1 (Ttg-1) encodes a nuclear protein normally expressed in neural lineage cells. Blood 77: 599–606

McGuire EA, Hockett RD, Pollock KM, Bartholdi MF, O'Brien SJ and Korsmeyer SJ (1989) The t(11; 14) (p15;q11) in a T-cell acute lymphoblastic leukemia cell line activates multiple transcripts, including Ttg-1, a gene encoding a potential zinc finger protein. Mol Cell Biol 9:2124–2132

McGuire EA, Rintoul CE, Sclar GM and Korsmeyer SJ (1992) Thymic overexpression of *Ttg-1* in transgenic mice results in T-cell acute lymphoblastic leukemia/lymphoma. Mol Cell Biol 12: 4186–4196

Michelsen JW, Scheichel KL, Beckerle MC and Winge DR (1993) The LIM motif defines a specific zinc-binding protein domain. Proc Natl Acad Sci USA 90: 4404–4408

Mouthon M-A, Bernard O, Mätjavila M-T, Romeo P-H, Vainchenker W and Mathieu-Mahul D (1993) Expression of *tal*-1 and GATA-binding proteins during human hematopoiesis. Blood 81: 647–655

Murre C, Bain G, van Dijk MA, Engel I, Furnari BA, Massari ME, Matthews JR, Quong MW, Rivera RR and Stuiver MH (1994) Structure and function of helix-loop-helix proteins. Biochim Biophys Acta 1218: 129–135

Murre C, McCaw PS and Baltimore D (1989a) A new DNA binding and dimerization motif in immunoglobulin enhancer binding, *daughterless, MyoD,* and *myc* proteins. Cell 56:777–783

Orkin SH (1992) GATA-binding transcription factors in hematopoietic cells. Blood 80: 575–581

Osada H, Grutz G, Axelson H, Forster A and Rabbitts TH (1995) Association of erythroid transcription factors: complexes involving the LIM protein RBTN2 and the zinc-finger protein GATA-1. Proc Natl Acad Sci USA 92: 9585–9589

Perez-Alvarado GC, Miles C, Michelsen JW, Louis HA, Winge DR, Beckerle MC and Summers MF (1994) Structure of the carboxy-terminal LIM domain from the cysteine rich protein CRP. Nature Struct Biol 1: 388–398

Pevny L, Simon MC, Robertson E, Klein WH, Tsai S-F, D'Agati V, Orkin SH and Costantini F (1991) Erythroid differentiation in chimeric mice blocked by a targeted mutation in the gene for transcription factor GATA-1. Nature 349: 257–260

Pui C-H (1995) Childhood Leukemias. New Eng J Med 332: 1618–1630

Pulford K, Lecointe N, Leroy-Viard K, Jones M, Mathieu-Mahul D and Mason DY (1995) Expression of TAL-1 proteins in human tissues. Blood 85: 675–684

Rabbitts TH (1994) Chromosomal translocations in human cancer. Nature 372: 143–149

Raimondi SC (1993) Current status of cytogenetic research in childhood acute lymphoblastic leukemia. Blood 81: 2237–2251

Royer-Pokora B, Loos U and Ludwig W-D (1991) TTG-2, a new gene encoding a cysteine-rich protein with the *LIM* motif, is overexpressed in acute T-cell leukaemia with the t(11; 14) (p13; q11). Oncogene 6: 1887–1893

Sadler I, Crawford AW, Michelsen JW and Beckerle MC (1992) Zyxin and cCRP: two interactive LIM domain proteins associated with the cytoskeleton. J Cell Biol 119: 1573–1587

Sanchez-Garcia I and Rabbitts TH (1993) LIM domain proteins in leukaemia and development. Sem Cancer Biol 4: 349–358

Sanchez-Garcia I and Rabbitts TH (1994) The LIM domain: a new structural motif found in zinc-finger-like proteins. Trends Genet 10: 315–320

Schmeichel KL and Beckerle MC (1994) The LIM domain is a modular protein-binding interface. Cell 79: 211–219

Shivdasani RA, Mayer EL and Orkin SH (1995) Absence of blood formation in mice lacking the T-cell leukaemia oncoprotein tal-1/SCL. Nature 373: 432–434

Valge-Archer VE, Osada H, Warren A, Forster A, Li J, Baer R and Rabbitts TH (1994) Two oncogenes RBTN2 and TAL1 involved in T cell acute leukaemias produce proteins which complex with each other in erythroid cells. Proc Natl Acad Sci USA 91: 8617–8621

Wadman I, Li J, Bash RO, Forster A, Osada H, Rabbitts TH and Baer R (1994) Specific in vivo association between the bHLH and LIM proteins implicated in human T cell leukemia. EMBO J 13: 4831–4839

Warren AJ, Colledge WH, Carlton MBL, Evans MJ, Smith AJH and Rabbitts TH (1994) The oncogenic cysteine-rich LIM domain protein Rbtn2 is essential for erythroid development. Cell 78: 45–57

Way JC and Chalfie M (1988) *mec*-3, a homeobox-containing gene that specifies differentiation of the touch receptor neurons in C. elegans. Cell 54: 5–16

Xia Y, Brown L, Yang CY-C, Tsan JT, Siciliano MJ, Espinosa III R, Le Beau MM and Baer RJ (1991) *TAL2*, a helix-loop-helix gene activated by the (7; 9) (q34; q32) translocation in human T-cell leukemia. Proc Natl Acad Sci USA 88: 11416–11420

The *TEL* Gene Contributes to the Pathogenesis of Myeloid and Lymphoid Leukemias by Diverse Molecular Genetic Mechanisms

T.R. Golub, G.F. Barker, K. Stegmaier, and D.G. Gilliland

1 Introduction

The *TEL* gene, originally cloned by virtue of involvement in the t(5;12) chromosomal translocation associated with chronic myelomonocytic leukemia (CMML), has a remarkable capacity to contribute to the pathogenesis of human leukemias: (1) *TEL* has been implicated in both myeloid and lymphoid leukemias, acute and chronic leukemias, and leukemias of both pediatric and adult populations; (2) *TEL*

Howard Hughes Medical Institute and the Division of Hematology/Oncology, Department of Medicine, Brigham and Women's Hospital, Harvard Medical School, Boston, MA 02115, USA

can contribute either its DNA binding domain or a putative helix-loop-helix (HLH) domain to fusion proteins, and (3) *TEL* has been associated with a surprising variety of fusion partners in human leukemias, including genes for transcription factors, receptor and non-receptor tyrosine kinases, and putative transcriptional activating domains. In addition, recent evidence suggests that loss of function of TEL may also contribute to pathogenesis of malignancy. In this report, the diverse molecular genetic mechanisms of leukemogenesis mediated by the *TEL* gene will be discussed.

2 Structure of the *TEL* Gene

TEL (for translocation, ets, leukemia) is a predicted 452 amino acid protein which has significant homology to the Ets family of transcription factors. The DNA-binding domain which defines Ets family members is located at the 3′ end of the *TEL* gene (Golub et al. 1994). In addition, TEL is one of the approximately one third of Ets family members which has a 5′ putative HLH domain (Fig. 1). The function of the HLH domain in Ets proteins is unknown, although deletion of the domain results in decreased transcriptional activation activity. The HLH domain is highly conserved, even in Ets family members in *Drosophila* such as *yan/pok*, and has weak homology with basic HLH (bHLH) domains in transcriptionally active

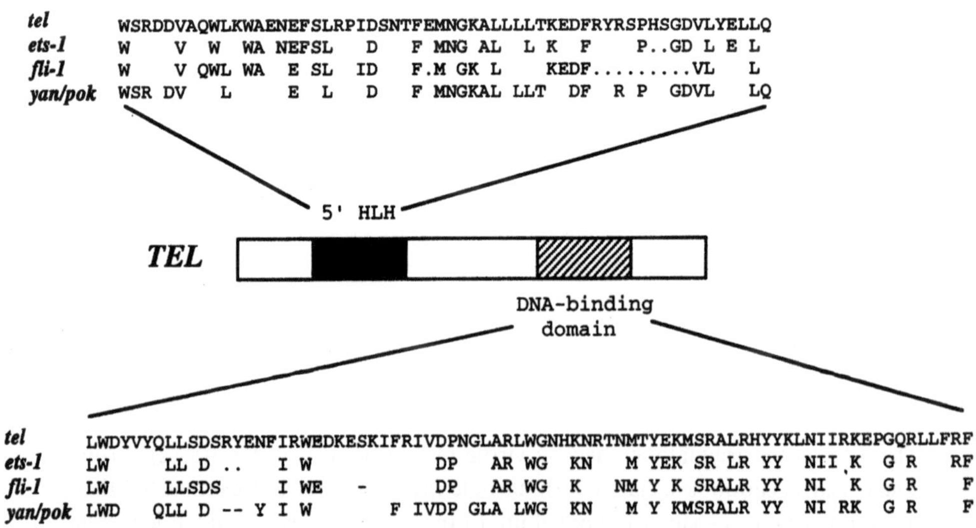

Fig. 1. The structure of the *TEL* gene. There is significant homology between *TEL* and other ets family members, including the mammalian *ets-1* and *fli-1* genes, as well as the *Drosophila* gene *yan/pok*. The region of homology on the 5' end of the gene has weak homology with helix-loop-helix motifs and may mediate protein-protein interactions. The DNA-binding domain defines this class of transcription factors and is also highly conserved

proteins such as myoD and myc. It is tempting to speculate that the putative HLH domain is a protein-protein interaction motif, although such a function has not been demonstrated for Ets proteins.

3 The *TEL* Helix-Loop-Helix Domain Is Fused to the Platelet-Derived Growth Factor Receptor-β Transmembrane and Tyrosine Kinase Domains in Chronic Myelomonocytic Leukemia Associated with t(5;12)

CMML is one of the French-American-British (FAB) subtypes of myelodysplastic syndrome (MDS), is characterized by dysplastic proliferation of monocytes, and progression to acute myeloid leukemia (AML). A recurring translocation in CMML between chromosome bands 5q33 and 12p13. t(5;12)(q33;p13) is of particular interest because it occurs in regions of chromosome 5q and 12p13 which are abnormal in a significant number of patients with hematologic malignancy. For example, approximately 10% of cases of acute lymphoblastic leukemia (ALL) of childhood are associated with 12p13 deletions (Pui et al. 1992; Raimondi et al. 1986). Since ALL is the most common cancer of children, (del)(12p) is perhaps the most frequent cytogenetic abnormality in pediatric malignancy. The t(5;12)(q33;12p13) may therefore serve to localize genes involved in CMML, as well as in other hematologic malignancies.

3.1 Cloning the t(5;12) Breakpoint Associated with CMML

We identified a patient with t(5;12)(q33;p13) and CMML who subsequently developed AML associated with acquisition of t(8;21)(q22:q22) in addition to t(5;12). The t(8;21) is identical at the cytogenetic level to t(8;21) seen in de novo AML. t(5;12) thus appears to satisfy criteria for an early molecular genetic abnormality which gives rise to AML by virtue of its recurring association with CMML. In this specific example, t(5;12) appears to be an early mutation antedating development of t(8;21) associated AML.

The t(5;12) breakpoint was cloned with limited clinical material (Golub et al. 1994). Fluorescence in situ hybridization (FISH) was performed with ordered chromosome 5q cosmid probes to localize the breakpoint between the c-*fms* and ribosomal protein *S14* genes. PCR primers specific for c-*fms* and RP*S14* were simultaneously used to screen the CEPH mega YAC library, and identified a 600 kb YAC 745d10 containing both c-*fms* and RP*S14*. 964c10 spanned the translocation by FISH, confirming the localization of the breakpoint and delineating a 600 kb genomic region within which the breakpoint must lie. Long range genomic maps were prepared by pulsed-field gel electrophoresis and regional probes, allowing localization of the breakpoint within the platelet-derived growth factor β

(PDGFRβ) gene. Ribonuclease protection assays (RPAs) were then used to localize the breakpoint with *PDGFRβ* RNA using *PDGFRβ*-specific probes on patient bone marrow RNA. RNA analysis localized a partial transcript for the 3′ end of the *PDGFRβ* gene, beginning near the transmembrane domain of PDGFRβ and extending through the tyrosine kinase domains. Northern blot analysis of patient bone marrow using *PDGFRβ* probes showed a single 5 kb transcript, which was larger than would be predicted by a partial *PDGFRβ* transcript, consistent with a *PDGFRβ* fusion partner derived from chromosome 12p13.

3.2 The *PDGFRβ* Fusion Partner in t(5;12) CMML Is a Novel ETS-Like Gene, *TEL*

Anchored PCR on patient marrow RNA was performed using nested *PDGFRβ* primers to obtain a partial cDNA for the chromosome 12 fusion partner. Involvement of the partial cDNA in the t(5;12) translocation was confirmed by RPA of patient bone marrow, and the partial cDNA was used to screen a human K562 cDNA library to obtain a full length cDNA.

The *PDGFRβ* fusion partner is a novel ETS-like gene, *TEL*. TEL is a 452 amino acid protein which contains two functional domains: (1) a 3′ DNA binding domain which defines Ets family members, and (2) a 5′ predicted HLH domain which is shared by approximately one third of Ets family members (Fig. 1). The TEL HLH is conserved among other Ets family members, including Ets-1 and fli-1 and *Drosophila* yan/pok, a transcriptional repressor. The consequence of the t(5;12) translocation is a fusion transcript whose expression is driven by the *TEL* promoter, and results in fusion of the TEL HLH domain in frame to the PDGFRβ transmembrane and tyrosine kinase domains (Fig. 2).

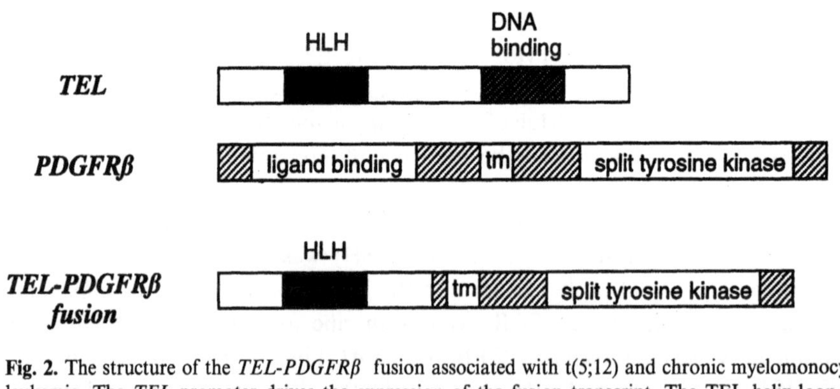

Fig. 2. The structure of the *TEL-PDGFRβ* fusion associated with t(5;12) and chronic myelomonocytic leukemia. The *TEL* promoter drives the expression of the fusion transcript. The TEL helix-loop-helix(HLH) domain is fused in frame to the transmembrane and tyrosine kinase domain of the Platelet-derived growth factor receptor - *β* (PDGFR*β*). The reciprocal transcript is not expressed. The TEL HLH domain may mediate dimerization and constitutive activation of the tyrosine kinase domain of TEL-PDGFR*β*

3.3 Mechanisms of Transformation of TEL-PDGFRβ

The structure of the TEL-PDGFRβ fusion product suggests several possible mechanisms of transformation. Wild-type PDGFRβ is known to signal a variety of cellular responses, including mitogenesis, on binding to its dimeric ligand, PDGF. PDGF mediates dimerization of PDGFRβ, which activates the tyrosine kinase leading to autophosphorylation of the receptor on tyrosine residues. Phosphorylated tyrosines on the PDGFRβ serve as docking sites for a number of proteins which initiate signal transduction cascades, including SRC, SYP/Grb2, PI3 kinase, and PLC_{γ}. It is plausible, based on the known function of wild-type PDGFRβ, that the HLH domain of TEL-PDGFRβ mediates dimerization and constitutive activation of the PDGFRβ tyrosine kinase domain.

The TEL-PDGFRβ fusion was first tested for transforming activity in cultured mammalian cell lines. TEL-PDGFRβ confers factor-independent growth to the interleukin-3 (IL-3)-dependent hematopoietic cell line, Ba/F3. Consistent with a model of PDGFRβ tyrosine kinase activation, TEL-PDGFRβ is constitutively phosphorylated in factor-independent Ba/F3 cells transfected with *TEL-PDGFRβ*. Deletion of the HLH domain, or inactivation of the tyrosine kinase domain, abrogates transforming activity of the TEL-PDGFRβ fusion. Efforts are currently underway in our laboratory to delineate which signal transduction pathways are required for transforming activity by TEL-PDGFRβ.

Another plausible model for TEL-PDGFRβ transforming activity is that TEL has tumor suppressor activity and that TEL-PDGFRβ interferes with wild-type function. In this model, the TEL HLH domain would mediate heterodimerization between TEL-PDGFRβ and wild-type TEL, leading to TEL loss of function. Indirect evidence which supports TEL loss of function in pathogenesis of malignancy is provided below and includes translocations involving *TEL* in which the other *TEL* allele is deleted, such as the *TEL-AML1* fusion. In these cases, there is no functional TEL in the leukemic cells: one *TEL* allele is deleted and the other is disrupted by translocation (Golub et al. 1995a). Other data supporting a role for TEL loss of function in hematologic malignancy is frequent loss of heterozygosity at the *TEL* gene locus in ALL (Stegmaier et al. 1995).

4 *TEL* Is Frequently Involved in Translocations at the 12p13 Locus

One rationale for cloning the t(5;12) translocation breakpoint was to determine whether the translocation would identify genes in other translocations involving chromosome 12p13. To test this possibility, additional patients with cytogenetic evidence of 12p13 rearrangements were analyzed for evidence of involvement of the *TEL* gene locus.

Yeast artificial chromosomes (YACs) containing the *TEL* gene were used to analyze patients with cytogenetic evidence of 12p13 abnormalities (Sato et al.

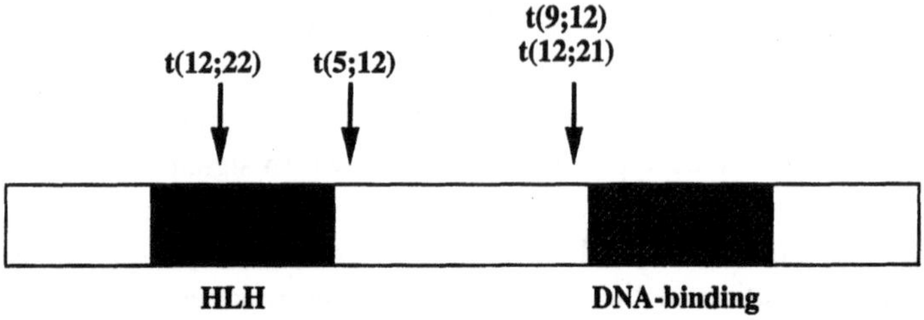

Fig. 3. Translocation breakpoints occur at diverse locations in the *TEL* gene. Ribonuclease protection analysis demonstrates diverse breakpoints within the *TEL* gene at 12p13 with various fusion partners on chromosomes 5q33, 9q34, 21q22, and 22q11

1995). The majority of patients (26/34) were shown to have abnormalities at the *TEL* gene locus. RPA was used to map to translocation breakpoints within *TEL* and disclosed unusual distribution of breakpoints within the *TEL* gene (Fig. 3). As noted earlier, translocation breakpoints within a given gene are usually highly conserved, even when different fusion partners have been identified. For example, the *MLL* gene at chromosome band 11q23 is associated with AML and has numerous fusion partners. However, the breakpoint with *MLL* is highly conserved, regardless of the fusion partner.

In contrast, there are at least three different breakpoints within the *TEL* gene which give rise to fusion products which express different functional domains of TEL. For example, in contrast with the structure of the TEL-PDGFRβ which involves the TEL HLH domain, patients evaluated in our laboratory with the t(12;22) showed evidence of abnormal expression of the TEL DNA binding domain driven by the promoter of a chromosome 22 gene. The t(12;22) breakpoint has been cloned by Grosveld and coworkers and gives rise to a fusion transcript derived from the *MNI* gene fused in frame to the DNA binding domain of the *TEL* gene (Fig. 4)(Buijs et al. 1995).

The MN1-TEL fusion is analogous in structure to the EWS-fli 1 fusion associated with t(11;22) in Ewings sarcoma (Sorensen et al. 1994) and the TLS-ERG fusion associated with t(16;21) leukemia (Shimizu et al. 1993), in which an Ets family DNA binding domain is aberrantly expressed.

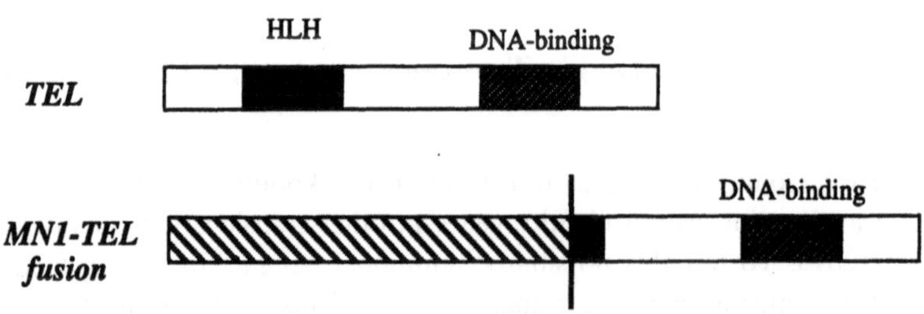

Fig. 4. t(12;22) fuses the *MN1* gene on chromosome 22q11 to the *TEL* gene DNA-binding domain. The *MN1-TEL* fusion is analogous to the *EWS-fli-1* fusion associated with t(11;22) and Ewing's sarcoma, and the *TLS-ERG* fusion associated with acute myeloid leukemia and t(16;21)

5 *TEL* Is Fused to the Proto-oncogene *ABL* in t(9;12;14) Acute Undifferentiated Leukemia

Ribonuclease protection assay was used as described above to delineate a breakpoint within the *TEL* gene in a patient with a complex t(9;12;14) translocation and acute undifferentiated leukemia with myeloid markers (AMoL). Anchored PCR with nested *TEL* primers was used to clone the *TEL* fusion partner, the *ABL* proto-oncogene on chromosome 9q34. The consequence of the translocation is fusion of the TEL HLH domain inframe to exon 2 of ABL (Fig. 5). The *TEL-ABL* fusion was recently cloned by Wiedemann and collaborators in a patient with pediatric ALL and t(9;12) (Papadopoulos et al. 1995; Golub et al. 1996).

The TEL-ABL fusion has several interesting features. *TEL-ABL* is similar in structure to the well characterized *BCR-ABL* fusion associated with chronic myelogenous leukemia (CML) and t(9;22). *TEL* is the only other fusion partner that has been identified for *ABL*, and has important similarities and differences from *BCR*. For example, *BCR* contains a 5′ predicted coiled-coil interaction motif which is necessary for tyrosine kinase and transforming activity of BCR-ABL. The coiled-coil motif of BCR and the putative HLH domain of TEL may both therefore serve dimerization or multimerization functions as a mechanism for constitutive activation of tyrosine kinase activity. The theme of dimerization leading to constitutive tyrosine kinase activity and transformation might then be shared by TEL-PDGFRβ, TEL-ABL and BCR-ABL. Consistent with this hypothesis, TEL-ABL transforms Rat1 fibroblasts and is constitutively phosphorylated when stably expressed in these cells. In addition, like TEL-PDGFRβ, TEL-ABL is capable of conferring IL-3-independent growth to Ba/F3 cells. Deletion of the HLH domain or inactivation of the tyrosine kinase of ABL abrogates transforming activity (Golub et al. 1996).

As another example of the usefulness of new fusion genes to elucidate functional domains which are relevant to transforming activity, TEL-ABL lacks a Grb2 binding site on the TEL moiety. BCR-ABL contains a Y177 Grb2 binding site on the BCR portion of the fusion, whose role in transformation has been debated

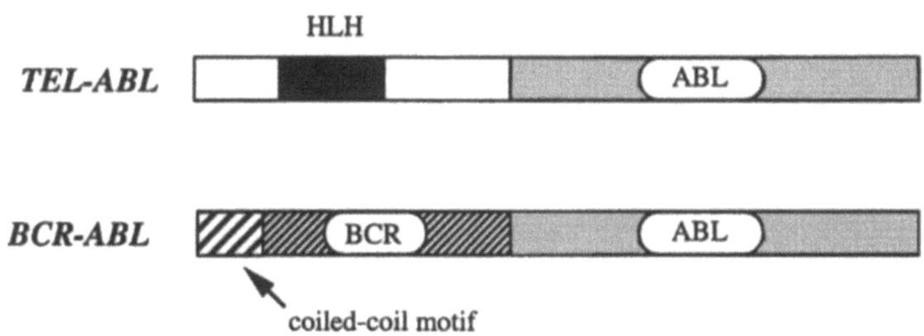

Fig. 5. The *TEL-ABL* fusion is analogous to the *BCR-ABL* fusion associated with t(9;22) chronic myelogenous leukemia. The TEL helix-loop-helix (HLH) domain is fused in frame to the ABL kinase. The TEL HLH domain is fused in frame to the ABL kinase. The TEL HLH domain and the coiled-coil motif in the BCR moiety may both serve to activate the ABL kinase by dimerization or multimerization

(Pendergast et al. 1993). Since TEL-ABL lacks a Grb2 binding site, at a minimum it can be stated that transformation of cultured mammalian cells mediated by ABL fusions does not require a functional Grb2 binding site on the ABL fusion partner.

6 *TEL* Is Fused to the *AML1* Gene in t(12;21) Associated with Pediatric Acute Lymphoblastic Leukemia

As noted above, another *TEL* breakpoint involving the TEL HLH domain was identified in patients with ALL and t(12;21). The translocation breakpoint was cloned using anchored PCR with *TEL*-specific primers. Based on our previous experience with cloning of *TEL-PDGFRβ* and *TEL-ABL*, one might have predicted a tyrosine kinase fusion partner for TEL. However, in the case of t(12;21), *TEL* is fused inframe to the transcription factor *AML1* (Golub et al. 1995a; Romana et al. 1995) (Fig. 6). The *AML1* gene on chromosome 21q22 was first cloned by virtue of its involvement with t(8;21) and t(3;21) associated with de novo AML and therapy-related AML, respectively (Nucifora et al. 1993). AML1 contains two functional domains; (1) a DNA binding domain with homology to the *Drosophila* pair-rule gene *runt*, and (2) a 3′ transcriptional activation domain. The TEL-AML1 fusion consists of the TEL HLH domain fused in frame to AML1 to intron 2, with expression of both the AML1 DNA binding domain and the AML1 transcriptional

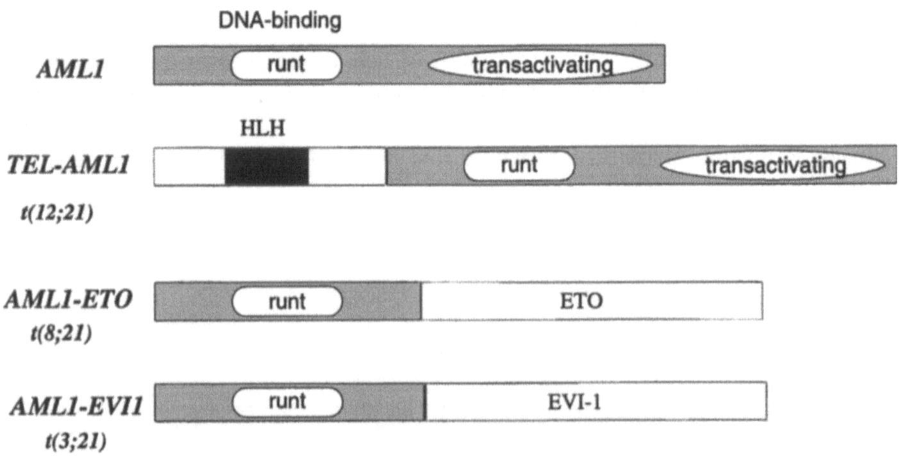

Fig. 6. The *TEL-AML1* fusion associated with t(12;21) and pediatric acute lymphoblastic leukemia (ALL). As noted in the text, this fusion is present in approximately 25% of cases of pediatric ALL, making it the most common gene rearrangement in childhood malignancy. The *TEL-AML1* fusion retains the transactivating domain of AML1, and thereby differs markedly from other fusions involving AML1. AML1 was named by virtue of association with myeloid leukemias; the *TEL-AML1* fusion demonstrates that AML1 can contribute to pathogenesis of lymphoid malignancy as well

activation domain. TEL-AML1 is fascinating from several perspectives. *First*, it suggests that the TEL HLH domain can contribute to pathogenesis of leukemia when fused either to a tyrosine kinase or to a transcription factor. *Second*, the structure of the *TEL-AML1* fusion differs significantly from the AML1 fusions involved in t(8;21) and t(3;21) translocations. In these translocations, the 5′ end of the *AML1* gene, including the DNA binding domain, is fused to one of several partners just 3′ of the *runt* domain (Fig. 6). Fusion partners include *ETO* in t(8;21), and various partners in t(3;21) including *EVI-1, EAP* and *MDS-1*. In each of these fusions, the AML1 trans-activation domain is lost. In contrast, in the *TEL-AML1* fusion, the full length *AML1* gene is expressed, including the *runt* and trans-activation domains. *Third*, AML1 had previously only been associated with myeloid leukemias (hence the name of the gene). Two cases of *TEL-AML1* reported from our laboratory and two cases subsequently reported by Romana et al. (1995) have been associated with lymphoid leukemias. In part, the difference in lineage specificity of *TEL-AML1* vs other *AML1* fusions can be explained by the t(8;21) and t(3:21) *AML1* fusions being driven by the *AML1* promoter, whereas the *TEL-AML1* fusion is driven by the *TEL* promoter. However, at a minimum it is clear that AML1 can contribute to the pathogenesis of both myeloid and lymphoid malignancies. *Fourth*, in each case of *TEL-AML1* fusions characterized this far (Golub et al. 1995a; Romana et al. 1995), the other *TEL* allele is deleted. Thus, in these leukemic cells, there is no functional TEL: one *TEL* allele is deleted and the other is disrupted by translocation. Based in part on this observation, the possibility that TEL loss of function might contribute to pathogenesis of leukemia was evaluated in ALL patients, as described in the next section.

7 Frequent Loss of Heterozygosity at the *TEL* Gene Locus in Pediatric Acute Lymphoblastic Leukemia

To evaluate the possibility that TEL loss of function might contribute to pathogenesis of leukemia, a patient population was chosen for analysis that has frequent deletions in the 12p13 region. Approximately 10% of pediatric ALL cases have 12p13 deletions. Since the most common childhood malignancy is ALL, del(12p13) is among the most common molecular genetic abnormalities of childhood cancer.

To determine whether the loss of TEL function could be implicated in pathogenesis of ALL, we first determined the frequency of loss of heterozygosity (LOH) at the *TEL* gene locus. Genomic DNA was prepared from 81 pediatric patients at the time of diagnosis of ALL. Polymorphic microsatellite markers D12589 and D12598 which flanked the *TEL* gene were then tested for LOH. LOH of two microsatellite markers which flank the *TEL* gene would provide convincing evidence for LOH at the *TEL* locus. As controls to confirm that ALL patients with a single microsatellite band had loss of heterozygosity, rather than simply being homozygous for that marker, paired leukemia and remission samples were ana-

Table 1. Loss of heterozygosity at the *TEL* gene locus in ALL patients

Marker	Patients (n)	Number informative	Percent with LOH
D12589	81	63/81 (78%)	9/63 (14%)
D12598	81	53/51 (65%)	9/63 (17%)

lyzed. Patients were considered informative only when remission samples documented that the patient was heterozygous at that locus.

As seen in Table 1, approximately 15% of ALL patients had LOH at the *TEL* gene locus (STEGMAIER et al. 1995). Of note, only one of the nine patients with *TEL* LOH had cytogenetic evidence of 12p13 loss. This is consistent with most studies of LOH in malignancy, in which cytogenetic analysis underestimates LOH at most loci. Taken together, these findings suggest that *TEL* LOH may occur in as many as 20%–25% of ALL patients.

To further delineate the region of LOH on 12p13, additional microsatellite markers telomeric and centromeric to *TEL* were evaluated. As shown in Fig. 6, the region of LOH includes *TEL*, but extends to the centromere and also invariably includes the gene *KIPI*, encoding the protein p27. p27 is a cyclin-dependent kinase CDK inhibitor which regulates the G1/S transition in the cell cycle. Other CDK inhibitors, such as p15 and p16, have been strongly implicated in pathogenesis of cancer through loss of function. p27 is thus a superb candidate for loss of function in ALL.

8 Loss of Heterozygosity Is Associated with *TEL-AML1* Fusion: The Most Common Gene Rearrangement in Childhood Acute Lymphoblastic Leukemia

We have recently obtained evidence which clarifies the finding of LOH at the *TEL* gene locus on 12p13 in pediatric ALL. As noted above, the t(12;21) associated with pediatric ALL results in fusion of *TEL* to *AML1*. In each of the cases described in the literature thus far, the *TEL-AML1* fusion has been associated with loss of the other allele of *TEL*. In addition, as noted above, the t(12;21) is a cryptic translocation: in most cases, there is no evidence at the cytogenetic level for the translocation. These considerations led us to explore the possibility that the frequency of t(12;21) translocations was higher than previously suspected. We first performed Southern blot analysis of the nine LOH patients described above and documented that, *in addition to deletion of one TEL allele, the other TEL allele was rearranged.* Furthermore, in patients for whom RNA was available, a *TEL-AML1* fusion could be documented. These findings suggest that LOH in this patient population was a marker for the *TEL-AML1* fusion. We then analyzed unselected pediatric ALL

Table 2. TEL contributes to the pathogenesis of human leukemias by diverse molecular genetic mechanisms

TEL involvement	Cytogenetic abnormality	Clinical phenotype
TEL HLH + tyrosine kinase		
TEL-PDGFRβ	t(5;12)	CMML
TEL-ABL	t(9;12)	AML, ALL
TEL HLH + transcription factor		
TEL-AML1	t(12;21)	25% of pediatric ALL; B-lineage
MN1 + DNA binding domain		
MN1-TEL	t(12;22)	AML
TEL translocation + TEL deletion		
TEL loss of function	t(9;12), t(12;21)	AML, ALL

HLH, helix-loop-helix; CMML, chronic myelomonocytic leukemia; PDGFR, platelet-derived growth factor receptor; AML, acute myeloid leukemia; ALL, acute lymphoblastic leukemia

cases for evidence of the *TEL-AML1* fusion. Of 42 pediatric patients with ALL, 10 (24%) were found to have *TEL-AML1* fusions by RT-PCR (GOLUB et al. 1995b).

To put these findings into perspective, the most common cytogenetic abnormality in pediatric ALL is t(1;19), associated with *E2A-PBX* fusion, and accounts for 5%–6% of cases. The cryptic t(12;21) which gives rise to the *TEL-AML1* fusion would therefore appear to be the most common molecular genetic abnormality in any childhood malignancy.

The reason for the association of the *TEL-AML1* fusion with frequent, if not invariant, loss of function of the other *TEL* allele is unclear. Since the TEL HLH domain may serve as a homodimerization motif, one possibility is that the residual *TEL* allele would be capable of interfering with the oncogenic potential of the *TEL-AML1* fusion. If this hypothesis is correct, this would represent a novel mechanism of carcinogenesis related to loss of function. Our laboratory is currently in the process of investigating this hypothesis.

9 TEL Contributes to the Pathogenesis of Leukemia by Diverse Molecular Genetic Mechanisms

In summary, we have presented evidence that the *TEL* gene, which we first cloned in association with t(5;12) CMML (GOLUB et al. 1994), can contribute to pathogenesis of leukemia by remarkably diverse mechanisms. The TEL HLH domain may be fused to the tyrosine kinase domains of PDGFRβ and ABL in myeloid leukemias, or may be fused to the transcription factor AML1 in lymphoid malignancies. In contrast, the TEL DNA binding domain may be aberrantly expressed in t(12;22) leukemias in a manner analogous to other Ets DNA binding domain fusions, such as EWS-fli1 and TLS-ERG.

Perhaps the most fascinating finding has been the frequent involvement of the *TEL-AML1* fusion in pediatric ALL. Characterization of the biological properties of TEL-AML1 should add insight to our knowledge of the molecular pathogenesis of ALL and may be useful in diagnosis and monitoring response to therapy.

The diversity of molecular genetic mechanisms by which TEL can be transforming suggests an important role for TEL in cell growth and differentiation (Table 2). Further analysis of TEL and its related oncogneic fusion genes, may provide further insight into the role of TEL in normal physiology of mammalian cells.

References

Buijs A, Sherr S, van Baal S, van Bezouw S, van der Plas D, van Kessel AG, Riegman P, Deprez RL, Zwartoff E, Hagemeijer A, Grosveld G (1995) Translocation (12;22)(p13;q11) in myeloproliferative disorders results in fusion of the ETS-like TEL gene on 12p13 to the MN1 gene on 22q11. Oncogene 10: 1511–1519

Golub TR, Barker GF, Lovett M, Gilliland DG (1994) Fusion of PDGF receptor beta to a novel ets-like gene, tel, in chronic myelomonocytic leukemia with t(5;12) chromosomal translocation. Cell 77: 307–316

Golub T, Barker GF, Bohlander S, Hiebert SW, Ward DC, Bray-Ward P, Morgan E, Raimondi SC, Rowley JD, Gilliland DG (1995a) Fusion of the TEL gene on 12p13 to the AML1 gene on 12q22 in acute lymphoblastic leukemia. Proc Natl Acad Sci 92: 4917–4921

Golub T, McLean T, Stegmaier K, Ritz J, Sallan S, Neuberg D, Gilliland DG (1995b) TEL-AML1: The most common gene rearrangement in childhood ALL. Blood (Suppl)

Golub TR, Goga A, Barker GF, Afar D, McLaughlin J, Bohlander SK, Rowley JD, Witte ON, Gilliland DG (1996) Oligomerization of the ABL tyrosine kinase by the ETS protein TEL in human leukemia. Mol Cell Biol (in press)

Nucifora G, Begy CR, Erickson P, Drabkin HA, Rowley JD (1993) The 3;21 translocation in myelodysplasia results in a fusion transcript between the AML1 gene and the gene for EAO, a highly conserved protein associated with the Epstein-Barr virus small RNA EBER 1. Proc Natl Acad Sci USA 90: 7784–7788

Papadopoulos P, Ridge SA, Boucher CA, Stocking C, Wiedemann LM (1995) The novel activation of ABL by fusion to an ets-related gene, TEL. Cancer Res 55: 34–38

Pendergast AM, Quilliam LA, Cripe LD, Bessing CH, Dai Z, Li N, Der CJ, Sclessinger J, Gishizky ML (1993) BCR-ABL-induced oncogenesis is mediated by direct interaction with the SH2 domain of the GRB-2 adapter protein Cell 75: 175–185

Pui CH, Raimondi SC, Crist WM (1992) Chromosomal abnormalities in childhood acute lymphoblastic leukaemia. Recent Adv Haematol 6: 89–105

Raimondi SC, Williams DL, Callihan T, Peiper S, Rivera GK, Murphy SB (1986) Nonrandom involvement of the 12p12 breakpoint in chromosome abnormalities of childhood acute lymphoblastic leukemia. Blood 68: 69–75

Romana SP, Mauchauffe M, LeConiat M, LePaslier D, Berger R, Bernard OA (1995) The t(12;21) of acute lymphoblastic leukemia results in a TEL-AML1 gene fusion. Blood 85: 3662–3670

Sato Y, Suto Y, Pietenpol J, Golub TR, Gilliland DG, Davis EM, LeBeau MM, Roberts J, Vogelstein B, Rowley JD, Bohlander SK (1995) TEL and KIP1 define the smallest region of deletions on 12p13 in hematopoeitic malignancies. Blood 86: 1525–1533

Shimizu K, Ichikawa H, Tojo A, Kaneko Y, Maseki N, Hayashi Y, Ohira M, Asano S, Ohki M (1993) An *ets*-related gene, ERG, is rearranged in human myeloid leukemia with t(16;21) chromosomal translocation. Proc Natl Acad Sci USA 90: 10280–10284

Sorensen PHB, Lessnick AL, Lopez-Terrada D, Liu XF, Triche TJ, Denny CT (1994) A second Ewing's sarcoma translocation, t(21;22), fuses the EWS gene to another ETS-family transcription factor, ERG. Nature Genetics 6: 146–151

Stegmaier K, Pendse S, Barker GF, Bray-Ward P, Ward DC, Montgomery KT, Krauter K, Reynolds C, Sklar J, Donnelly M, Bohlander SK, Rowley JD, Sallan SE, Gilliland DG, Golub TR (1995) Frequent loos of heterzygosity at the TEL gene locus in acute lymphoblastic leukemia of childhood. Blood 86: 38–44

Characterisation of the *PML/RARα* Rearrangement Associated with t(15;17) Acute Promyelocytic Leukaemia

D. Grimwade and E. Solomon

1 Introduction

Acute promyelocytic leukaemia (APL; FAB AML M3, Bennett et al. 1976) represents a unique example of a disease in which a successful treatment approach, in the form of all-*trans* retinoic acid (ATRA), has been developed that directly addresses and overcomes the causative molecular abnormality. For over a decade, retinoids have been noted to possess therapeutic activity which is virtually specific to the acute promyelocytic form of acute myeloid leukaemia (AML) (Breitman et al. 1981). Subsequent clinical trials have shown that ATRA can achieve remission rates of over 90% in APL (Huang et al. 1988; Castaigne et al. 1990; Chomienne et al. 1990), representing an apparent significant improvement on results obtained with conventional chemotherapy; indeed a number of patients in these studies were chemoresistant or treated in relapse. Parallel in vitro studies have demonstrated that remission is achieved by terminal differentiation of the leukaemic clone rather than by a cytotoxic effect (Huang et al. 1988; Castaigne et al. 1990; Chomienne

Somatic Cell Genetics Laboratory, Imperial Cancer Research Fund, 44, Lincoln's Inn Fields, London, WC2A 3PX, UK

et al. 1990). This has been confirmed using clonal analysis of APL blasts and peripheral blood neutrophils following ATRA therapy (Elliott et al. 1992).

Since the late 1970s it has been appreciated that APL is characterised by a reciprocal translocation, involving chromosomes 15 and 17, present in virtually all cases (Rowley et al. 1977). However, it was not until 1990 that the APL breakpoint region was ultimately cloned, and the significance of the response to retinoids became apparent (Borrow et al. 1990; de Thé et al. 1990; Lemons et al. 1990; Alcalay et al. 1991). The translocation was found to disrupt a previously uncharacterised gene, *PML*, on chromosome 15 and the retinoic acid receptor-α gene (*RARα*) on chromosome 17. RARα belongs to the steroid hormone receptor superfamily and acts as a transcription factor, mediating the effect of retinoic acid at specific response elements (Giguere et al. 1987; Leid et al. 1992a). The translocation leads to the formation of *PML-RARα* and *RARα-PML* chimaeric genes. *PML-RARα* transcripts derived from add(15q) are invariably present and are thought to mediate leukaemogenesis, whereas *RARα-PML* is transcribed in approximately 80% of cases (Alcalay et al. 1992; Grimwade et al. 1996b) and its significance remains unclear.

The realisation that the PML-RARα fusion protein retains the retinoic acid ligand binding domain suggested that it may not only mediate leukaemogenesis, but that it may in some way account for the unique sensitivity of APL to differentiation by retinoids such as 9-*cis* retinoic acid and ATRA. Therefore over the last few years research has been aimed at determining: (1) the normal function of PML, (2) the role of retinoid receptors in haematopoietic differentiation, (3) how the rearrangement of *PML* and *RARα* might promote leukaemogenesis and (4) the mechanisms by which ATRA reverses the leukaemic phenotype.

2 Clinical Features of Acute Promyelocytic Leukaemia

Acute promyelocytic leukaemia is one of the commoner forms of AML, accounting for approximately 10% of de novo cases, typically presenting in early middle age (Stone and Mayer 1990; Avvisati et al. 1992). The disease is believed to represent a clonal expansion of haematopoietic precursor cells associated with a differentiation block in myeloid development that leads to an accumulation of abnormal promyelocytes (Grignani et al. 1993). Two major morphological subtypes of APL can be distinguished : (1) "classical" M3 (Bennett et al. 1976) and (2) the M3 variant (M3v) (Golomb et al. 1980), although a much rarer hyperbasophilic form has also been described (McKenna et al. 1982). In the majority of patients, the marrow is replaced by hypergranular promyelocytes, which often have numerous Auer rods, so-called faggot cells (Bennett et al. 1976). Patients with classical APL typically present with pancytopaenia with scanty abnormal cells in the peripheral blood. Approximately a quarter of patients demonstrate the variant morphological form (Stone and Mayer 1990), in which leukaemic cells, particularly in the peripheral blood, appear less heavily granulated by light microscopy with conventional MGG staining and typically have bilobed nuclei (GOLOMB et al. 1980). Nevertheless, electron microscopy and immunocytochemistry reveal that the leukaemic cells still contain numerous granules, which are smaller than those of classical APL (GOLOMB et al. 1980); hence the term hypogranular or microgranular variant APL (M3v) has arisen. Patients with M3v often present with leucocytosis (AVVISATI et al. 1992) and the peripheral blood picture may be confused with that of myelomonocytic or monocytic/monoblastic leukaemias (AML M4/5). These distinct entities may be readily distinguished by immunocytochemistry, immunophenotyping and cytogenetic assessment (BAIN 1990), without necessarily resorting to molecular diagnostic techniques. The immunophenotypes of classical APL and M3v are broadly similar, being characterised by the presence of CD9, CD13, CD33 and sialylated CD15, with poor expression of HLA-DR (AVVISATI et al. 1992; PAIETTA et al. 1994; DI NOTO et al. 1994; VAHDAT et al. 1994). It has recently been appreciated that patients with M3v may aberrantly express the T cell antigen, CD2 (BIONDI et al. 1993; 1995). The division of APL into classical and variant morphological subgroups may be somewhat artificial, since they are probably extreme ends of a spectrum. Furthermore, the morphological appearances may be interchangeable, since patients with classical APL may demonstrate reduced granularity on relapse, whereas culture of variant cells may be associated with an increase in granularity more typical of classical morphology (CASTOLDI et al. 1994).

The heavy granulation associated with APL cells has been considered to underlie the severe haemorrhagic problems that typify the disease (LINCH et al. 1994). Patients often experience morbidity and mortality relating to a bleeding diathesis that is more severe than would be predicted from the level of thrombocytopaenia alone (LINCH et al. 1994). This is believed to reflect some degree of activation of the coagulation cascade and more importantly increased fibrinolytic activity due to release of plasminogen activators and proteolytic enzymes (AVVISATI et al. 1992; TALLMAN and KWAAN 1992). The coagulopathy may be exacerbated on commencement of chemotherapy, presumably due to further release of procoagulant and fibrinolytic factors as APL cells become disrupted (TALLMAN and KWAAN 1992). ATRA, associated with remission induction by differentiation rather than cell lysis, has been found to ameliorate the coagulopathy (CASTAIGNE et al. 1990), leading to a rapid improvement in hyperfibrinolysis/proteolysis, although the excess procoagulant activity leading to thrombin generation may be more persistent (RODEGHIERO and CASTAMAN 1994).

Despite the risk of early mortality secondary to the bleeding diathesis, a number of studies using conventional chemotherapy have demonstrated that APL represents a relatively favourable prognostic group of AML (SWANSBURY et al. 1994; BURNETT 1994). This reflects both a reduced risk of relapse and often a good response to reinduction therapy should relapse occur. Remission induction achieved with ATRA, in contrast to chemotherapeutic agents, does not induce marrow aplasia, reflecting its differentiating effect on the leukaemic clone with a probable concomitant stimulation of normal colony forming activity (SAKASHITA et al. 1993). Marrow aplasia increases the risk of death from haemorrhage and infection. Therefore, it was hoped that initial treatment of APL with ATRA, by both ameliorating the coagulopathy and achieving remission without marrow aplasia, might lead to further improvements in complete remission rates and hence overall survival. However, it is now clear that early benefits of ATRA may be offset due to morbidity or mortality caused by a constellation of clinical features, known as the "ATRA syndrome", that develop in about 30% of patients (FRANKEL et al. 1992). This phenomenon, which curiously appears to be less common amongst patients treated in China, may be secondary to the modulation of surface markers as differentiation occurs (DI NOTO et al. 1994). ATRA syndrome may lead to the development of fluid retention associated with pleural and pericardial effusions and pulmonary infiltrates which can precipitate death due to respiratory failure (FRANKEL et al. 1992); although this may be averted if steroids are started promptly (VAHDAT et al. 1994). An even more significant problem with the use of ATRA as a single agent therapy is that it does not lead to the eradication of the disease-related clone: hence responses are not durable without consolidation chemotherapy. However, recent trials have demonstrated that treatment with ATRA and chemotherapy is far superior to chemotherapy alone (FENAUX et al. 1994). In a French trial, chemotherapy-treated patients had a disease free survival of 42% at 4 years as compared to 70% in patients receiving combined therapy (FENAUX et al. 1994). This has led to the development of trials such as the UK Medical Research Council ATRA study which seeks to address the most efficacious means of combining ATRA and chemotherapy.

3 Characterisation of the Acute Promyelocytic Leukaemia Breakpoint Regions

The APL breakpoint region was successfully characterised by employment of two distinct approaches namely, positional cloning and candidate gene screening. BORROW et al. (1990) used a *Not*1 linking library from an interspecies hybrid, containing 17q as its only human material, to identify *Not*1 sites in the region of the APL breakpoint. *Not*1 linking clones were generated which were found to flank the breakpoint and used as probes on pulsed field gel electrophoresis. A subclone was isolated which identified rearrangements on Southern blots made from somatic cell hybrids incorporating the APL 15q derivative chromosome and from material derived directly from APL patients. Ultimately, screening an HL60 cDNA library revealed that the gene rearranged on 17q was RARα, with disruption occurring within the second intron (numbering system of BRAND et al. 1990).

RARα was screened independently as a candidate gene by other groups, knowing its proximity to the APL breakpoint region on 17q and bearing in mind the unique sensitivity of the disease to retinoids such as ATRA (LONGO et al. 1990; DE THÉ et al. 1990; ALCALAY et al. 1991).

RARα was found to be fused to a novel gene on chromosome 15q, initially known as *myl* and subsequently renamed *PML* (DE THÉ et al. 1990, 1991; ALCALAY et al. 1991; GODDARD et al. 1991; KAKIZUKA et al. 1991). Characterisation of PML led to the identification of a number of motifs, initially suggesting a role as a putative transcription factor (Fig. 1) (KAKIZUKA et al. 1991; DE THÉ et al. 1991; GODDARD et al. 1991). The NH_2-terminal region of PML is proline-rich, as previously found in the transcriptional activation domain of CTF (MERMOD et al. 1989) and in a number of other transcription factors (MITCHELL and TJIAN 1989). Adjacent to this region are three cysteine-rich motifs, followed by a coiled-coil domain (KAKIZUKA et al. 1991; DE THÉ et al. 1991; GODDARD et al. 1991; FREEMONT 1993). The most NH_2-terminal of these domains has a C_3HC_4 configuration of cysteine and histidine residues which is homologous to the zinc finger motif first identified in RING 1 (FREEMONT et al. 1991). This motif, which has subsequently been identified in an ever-increasing number of proteins, including recently the predicted product of the *BRCA*1 breast cancer susceptibility gene (MIKI et al. 1994), has become known as the RING finger domain (FREEMONT 1993). PML belongs to a subgroup of RING finger containing proteins that are characterised by two further cysteine/histidine-rich regions with homology to the RING finger itself, known as "B-boxes", which are immediately adjacent to the α-helical coiled-coil domain (FREEMONT 1993). Towards the COOH-terminal of PML is a serine/proline-rich region which has been considered a potential site of phosphorylation by casein kinase II (KAKIZUKA et al. 1991; KASTNER et al. 1992).

By comparison of the predicted amino acid sequence of RARα with that of steroid and thyroid hormone receptors, six regions have been defined on the basis of sequence conservation (GIGUERE et al. 1987; LEID et al. 1992a). At the NH_2-terminal is the A/B domain which is considered to have a ligand-independent transactivation function ("AF-1") (LEID et al. 1992a), modulating and augmenting the relatively more important ligand-dependent trans-activation function ("AF-2") mediated by COOH-terminal regions of the receptor (NAGPAL et al. 1992). Studies using truncated retinoic acid receptors (RARs) have demonstrated that the functional activity attributable to the NH_2-terminal varies considerably depending upon the retinoid response element and promoter employed (NAGPAL et al. 1992). This suggests that the A/B domain may influence the pattern of retinoid responses, which is of interest considering that in APL RARα is disrupted between the A and B domains (BORROW et al. 1990). The adjacent C domain contains two zinc finger motifs which are required for DNA binding; the stem of the most NH_2-terminal finger confers specificity of the interaction, lying within the major groove of DNA (LUISI et al. 1991; SCHWABE et al. 1993). In steroid hormone and retinoid receptors, the adjacent D domain may provide a nuclear localisation signal (BEATO 1989). The E region, in addition to containing the ligand-binding domain and conserved amphipathic α-helical motif conferring AF-2 activity, demonstrates a series of hydrophobic heptad repeats which form an important dimerisation domain, mediating the interaction between RARs and retinoid-X receptors (RXRs) (BEATO 1989; LEID et al. 1992a,b; DANIELAN et al. 1992; AU-FLIEGNER et al. 1993; DURAND et al. 1994). The integrity of the ninth heptad repeat has been found to be essential;

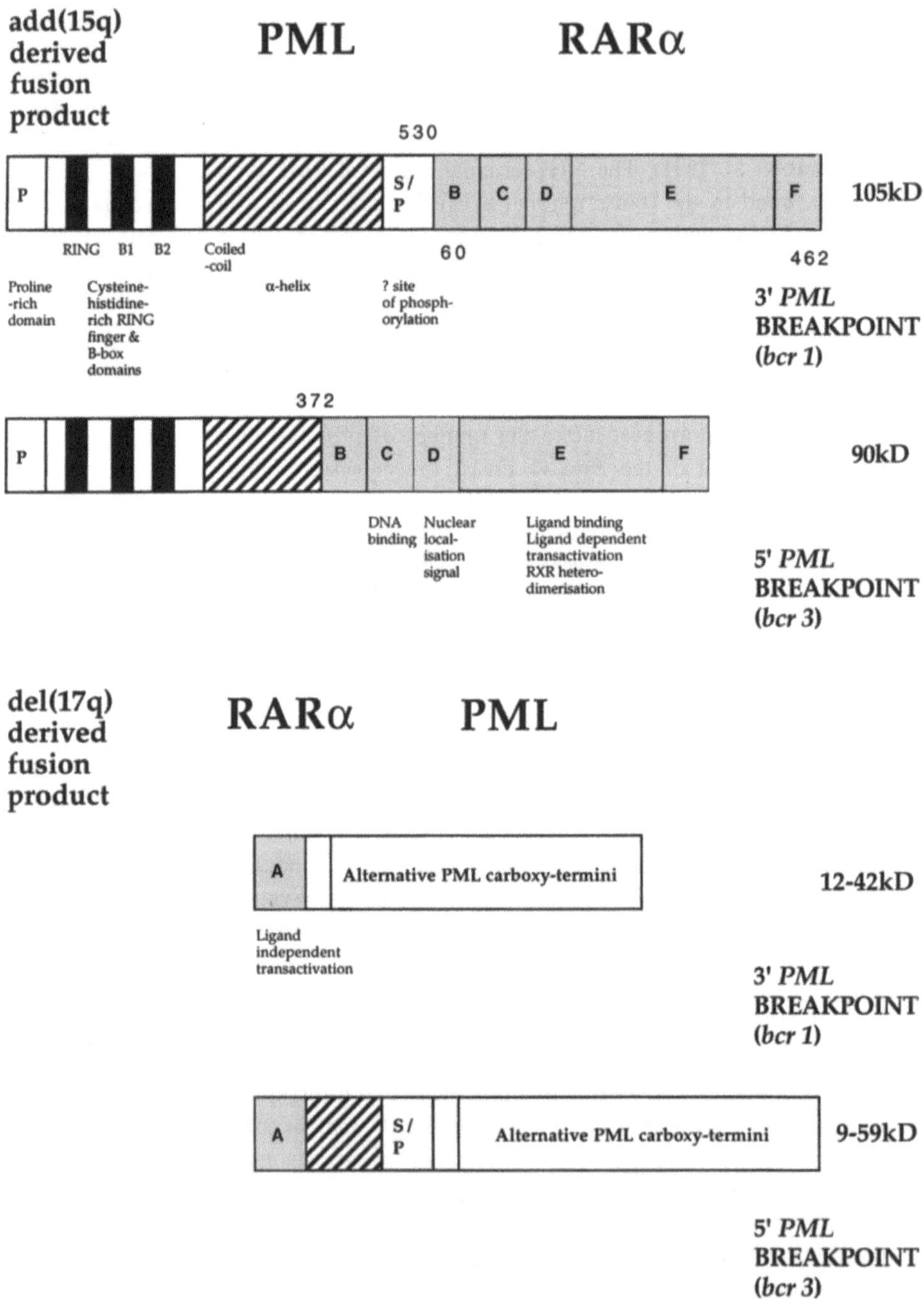

Fig. 1. The fusion products generated by t(15;17) in APL, associated with the commonest *PML* breakpoint sites (3′ = bcr1; 5′ = bcr 3) (From Goddard et al. 1991; Pandolfi et al 1992; Alcalay et al. 1992)

mutations of the first and eighth hydrophobic residues of this repeat abolish the RAR/RXR interaction, although in functional studies mutations of the first amino acid can be overcome by the presence of ligand (Au-Fliegner et al. 1993). The role of the COOH-terminal F domain of RARs is not clear, particularly as it is absent in the RXRs (Leid et al. 1992b).

The 15;17 translocation leads potentially to the formation of three different abnormal products: PML-RARα derived from add(15q), RARα-PML from del(17q) and aberrant truncated PML, (see Fig. 1) (de Thé et al. 1990; Goddard et al. 1991; de Thé et al. 1991; Kakizuka et al. 1991; Pandolfi et al. 1992; Borrow et al. 1992; Alcalay et al. 1992). The PML-RARα product involves fusion of the NH_2-terminal portion of PML, including the proline-rich region and tripartite motif of RING finger, B-boxes and coiled-coil to the B domain of RARα. Hence the fusion protein includes the DNA binding, ligand binding and heterodimerisation motifs of RARα. RARα-PML retains only the RARα trans-activation domain (A) and variable COOH-terminal portions of PML including potential phosphorylation sites (Alcalay et al. 1992; Borrow et al. 1992). Aberrant COOH-terminally truncated PML proteins, lacking either part of the α-helix or the whole α-helix and second B-box motif (B2), may be transcribed (Pandolfi et al. 1992; Fagioli et al. 1992). These could potentially disrupt wild-type PML function and possibly play a contributory role in leukaemogenesis. However since PML-RARα is invariably present, this is the most likely mediator of leukaemic transformation. The product retains all the functionally important domains that confer the activity of RARα, but transcription is no longer under the control of the RARα promoter. Furthermore the retention of RING, B-box and coiled-coil motifs of PML in the fusion protein is reminiscent of two other RING family members: TIFI (Miki et al. 1991; Le Douarin et al. 1995) and RFP (Takahashi et al. 1988), which are also capable of forming oncogenic fusion products (Kastner et al. 1992; Freemont 1993). This suggests that these motifs play an important role in the transformation potential of these proteins.

The APL breakpoint on chromosome 17 invariably occurs within the RARα second intron (Borrow et al. 1990; Pandolfi et al. 1992). Recently a study of three patients has suggested that some breakpoint cluster within a 50bp region associated with a high frequency of topoisomerase I and II sites. An in vitro transfection-recombination assay suggested that this region may represent a hot spot for illegitimate recombination (Tashiro et al. 1994). In contrast to RARα, two major breakpoint regions have been delineated within *PML*. In approximately a third of patients a 5′ breakpoint occurs (*bcr3*), usually within intron 3 (Pandolfi et al. 1992; Biondi et al. 1992), (Fig. 2). This leads to the formation of a fusion product comprising the NH_2-terminal 372 amino acids of PML linked to the COOH-terminal 403 residues of RARα (Goddard et al. 1991; Kakizuka et al. 1991), predicting a 90kDa protein (Fig. 1). In the remaining two thirds of cases the *PML* breakpoint occurs more 3′ (Biondi et al. 1992). In the majority of these patients and in the APL cell line NB4 (Lanotte et al. 1991) the breakpoint lies within intron 6 (*bcr1*), whereas in approximately 10% of patients disruption occurs within exon 6 (*bcr2*) (Pandolfi et al. 1992; Gallagher et al. 1995). The *bcr1* pattern leads to the

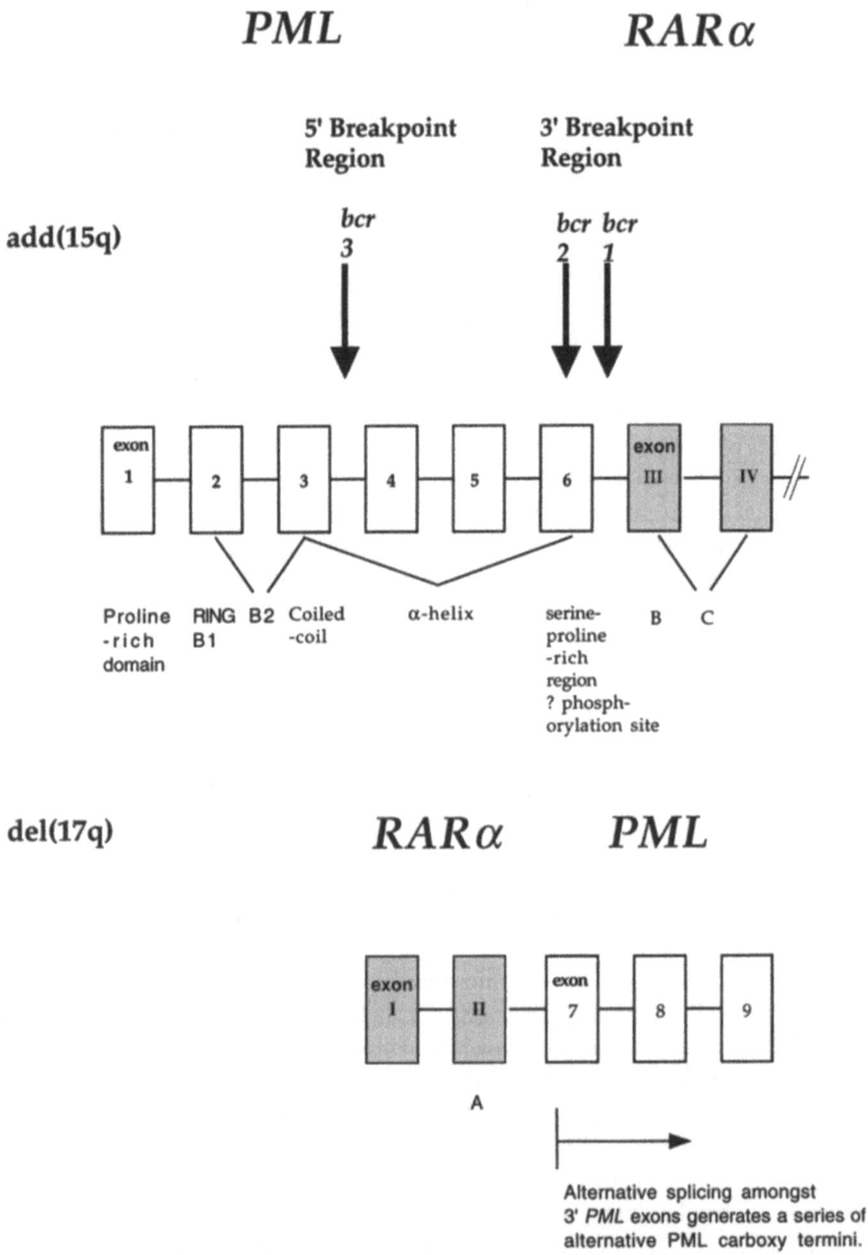

Fig. 2. The genomic organisation at the chromosome 15 and 17 breakpoint regions of a patient with a 3′ (*bcr* 1) *PML* breakpoint. The positions of other potential *PML* breakpoints (*bcr* 2 & 3) are denoted. *Bcr 3* breakpoints lead to the translocation of *PML* exons 4–9 to chromosome 17. In patients with a *bcr 2* breakpoint the *RARα-PML* reciprocal fusion gene includes variable portions of *PML* exon 6 in addition to exons 7–9. Exons 4–9 of *PML* are subject to alternative splicing, generating multiple PML-RARα & RARα-PML fusion products and up to 13 PML isoforms. (From Goddard et al. 1991; Pandolfi et al. 1992; Alcalay et al. 1992; Borrow et al. 1992)

fusion of the NH_2-terminal 530 aminoacids of PML to RARα predicting a protein of 105 kDa (Goddard et al. 1991, Pandolfi et al. 1991; 1992). Patients with a *bcr2* pattern usually have a fusion protein of 96–105 kDa, the size being partly determined by the breakpoint position within exon 6 (Pandolfi et al. 1992). A number of studies have demonstrated no significant correlation between *PML* breakpoint pattern and various disease parameters such as presence of coagulopathy or variant morphology (Borrow et al. 1992; Fukutani et al. 1995). However, a trend towards increased frequency of *bcr*3 breakpoints in patients with M3v has been noted in some series (Biondi et al. 1992; 1994) and this may explain the contentious association with CD2 expression (Claxton et al. 1992; Biondi et al. 1993, 1995; Maslak et al. 1993).

PML-RARα fusion products are subject to considerable heterogeneity generated not only by three possible *PML* breakpoints, but also due to alternative splicing involving exons 3,4,5 and 6 of *PML* in addition to the use of two alternative RARα polyadenylation region (Pandolfi et al. 1992), (Fig. 2). *PML* exons 3-5 encode the α-helix, the most NH_2-terminal part of which forms the coiled-coil domain, exon 6 corresponds to part of the COOH-terminal serine/proline-rich potential phosphorylation region and may also encode a nuclear localisation signal (Fagioli et al. 1992; Kastner et al. 1992). Therefore, the *bcr*3 breakpoint generates fusion products lacking the COOH-terminal phosphorylation region associated with some reduction in α-helix length; products in which exon 3 is spliced out lack the coiled-coil domain altogether (Pandolfi et al. 1992) (Fig. 1). Patients with a *bcr1* pattern generate a series of fusion products which contain a longer PML α-helix than in *bcr3* and may retain part of the COOH-terminal phosphorylation region. The domains present in *bcr2* patients depend upon the position of the exon 6 breakpoint. The alternative splicing of exons 3-6 noted in patients with 3′ *PML* breakpoints is also present in the reciprocal *RARα-PML* transcripts of patients with 5′ (*bcr3*) breakpoints (Borrow et al. 1992).

This heterogeneity in fusion products generated by t(15;17) creates confusion as to their relative roles in mediating leukaemogenesis and raises concerns regarding the selection of appropriate constructs for functional studies. However, since the majority of PML-RARα fusion products retain the proline-rich, RING-finger, B-boxes and the coiled-coil motif, there may be little functional difference between them with regard to the ability to cause leukaemic transformation. More subtle differences between fusion products, for example altered length of α-helix, the variable retention of domains containing potential sites of phosphorylation or conferring nuclear localisation and whether there is concomitant expression of RARα-PML might account for any differences in disease characteristics of patients with different breakpoint patterns. In this regard it has recently been noted that patients with a *bcr3* pattern have a worse prognosis, associated with an increased risk of relapse (Vahdat et al. 1994), although this remains contentious (Fukutani et al. 1995).

4 The Wild-Type PML Protein

The discovery of the *PML-RARα* rearrangement in APL paved the way for further characterisation of wild-type PML, the precise function of which still remains unclear. It is expressed in a wide range of tissues and haematopoietic cell lines as a bewildering array of isoforms that range in molecular weight from 48 to 98 kDa (GODDARD et al. 1991; FAGIOLI et al. 1992; BORROW and SOLOMON 1992; TERRIS et al. 1995; FLENGHI et al. 1995). Isoform variability is generated by alternative splicing within the central exons (4,5 and 6) in addition to the utilisation of a series of potential COOH-terminals (exons 7, 8 and 9) (GODDARD et al. 1991; FAGIOLI et al. 1992; BORROW and SOLOMON 1992). The significance of this variability is unclear, particularly as all isoforms retain the cysteine-rich motifs and coiled-coil domains that are considered of functional importance (GODDARD et al. 1991; FAGIOLI et al. 1992). Alternative splicing between the central exons could influence the distribution of PML within the cell, by an effect on the nuclear localisation signal (KASTNER et al. 1992; FLENGHI et al. 1995). It has also been suggested that COOH-terminal variability might alter the biological activity of PML by inserting regions with potential phosphorylation sites that could alter the function of the serine/proline-rich domain (FAGIOLI et al. 1992).

The presence of the NH_2-terminal proline-rich region, three zinc finger motifs and the adjacent coiled-coil domain with homology to the Jun/Fos dimerisation interface suggested to many groups that PML is a putative transcription factor whose function is disrupted in APL (GODDARD et al. 1991; KAKIZUKA et al. 1991; DE THÉ et al. 1991; KASTNER et al. 1992). However, further study of many RING-finger proteins including PML has revealed no direct evidence for specific DNA binding (FREEMONT 1993). The RING finger proteins represent a large group with a variety of regulatory functions including a number involved in development and cell differentiation (FREEMONT, 1993). In proteins such as yeast PEP3 (PRESTON et al. 1991; ROBINSON et al. 1991) and PEP5 (DULIC and RIEZMAN 1989; WOOLFORD et al. 1990), which belong to the RING H2 family and are involved in vacuolar biogenesis, the RING-finger domain is clearly not involved in an interaction with either DNA or RNA and is a site of protein-protein or protein-lipid interaction (FREEMONT 1993). The function of other RING-finger proteins more clearly implies a close relationship with DNA processing. RAG1 is involved in the activation of V(D)J recombination in lymphoid cells (SCHATZ et al. 1989), RAD18 is required for postreplicative repair of UV damaged DNA in yeast (JONES et al. 1988) and TIFI is a putative mediator of ligand-dependent transactivation of steroid hormone nuclear receptors (LE DOUARIN et al. 1995). It is possible that these effects are not mediated by direct DNA binding of the RING finger itself, but may involve the formation of complexes with other proteins that are responsible for binding to DNA. Hence, RING finger proteins could influence transcription rates by dimerising with transcriptional activators and/or repressors and possibly form complexes with basal transcription machinery. In this regard the RING finger and B-box domains could represent potential sites of protein-protein interaction. The homology of the coiled-coil domain of PML to the Jun/Fos dimerisation motif suggests a further potential site of association with transcription factors and may account for the apparent interaction between PML-RARα and AP1 signalling (DOUCAS et al. 1993). However, there have been no reports investigating any interplay between wild-type PML and AP1 responsive pathways.

More insight into the normal function of PML has been provided by the development of PML antisera, which have characterised its intracellular distribution. PML has been found to localise predominantly to discrete structures within the nucleus, known variably as nuclear bodies, ND10, Kr bodies and PODs (STUURMAN et al. 1992; KASTNER et al. 1992; DANIEL et al. 1993; DYCK et al. 1994; WEIS et al. 1994; KOKEN et al. 1994; BORDEN et al. 1995), although cytoplasmic aggregates of PML may also be detected in 10%–20% of cells (FLENGHI et al. 1995). The nuclear bodies are composed of several proteins, including NDP52 (KORIOTH et al. 1995; PIC 1, a novel ubiquitin-like protein BODDY et al. 1996) and SP100, a 75 kDa autoantigen recognised by anti-Kr sera (WEIS et al. 1994). Antibodies to SP100 may be detected in some patients with autoimmune diseases, particularly primary biliary cirrhosis (WEIS et al. 1994). The function of these nuclear bodies is ill-defined at present: there are usually 10–30 per nucleus, they lie in a close relationship to chromatin, are distinct from nucleoli and spliceosomes (WEIS et al. 1994) and do not associate with metaphase chromosomes (FLENGHI et al. 1995).

Recent studies have demonstrated that the PML nuclear bodies form a target for a variety of viral infections, including herpes simplex (EVERETT and MAUL 1994), adenovirus (PUVION-DUTEILLEUL et al. 1995; CARVALHO et al. 1995; DOUCAS et al. 1996), and cytomegalovirus (KELLY et al. 1995). For example, in the case of herpes simplex, Vmw110, an immediate early viral protein required for reactivation of latent virus and also incidentally a RING finger protein, colocalizes with PML nuclear bodies early in the course of infection (EVERETT and MAUL 1994). With subsequent progression, Vmw110 and PML are ultimately translocated together to the cytoplasm (EVERETT and MAUL 1994), although it is unclear at present whether other constituents of the nuclear domain are also transported in this process. However, this phenomenon has led to the suggestion that the PML-containing nuclear bodies may mediate a storage function or possibly be involved in trafficking between nucleus and cytoplasm (EVERETT and MAUL 1994). This redistribution of PML in viral infection is interesting, in the context of recent studies demonstrating marked expression of PML in macrophages, particularly associated with various inflammatory and neoplastic processes (FLENGHI et al. 1995; TERRIS et al. 1995). PML expression appears to be upregulated in neoplastic cells of a wide range of tumours, in cells affected by various inflammatory processes, in normal tissues associated with proliferation and in promonocytic cells induced to differentiate with combinations of vitamin D_3 and transforming growth factor (TGF)β1 with or without interferon-γ (FLENGHI et al. 1995; TERRIS et al. 1995). Whether these findings are of any functional significance, perhaps indicating that PML is a marker of the proliferative state of the cell and possibly implying that it is involved in cell growth, or whether they merely reflect a nonspecific cytokine mediated effect is unclear at present. However, it is now apparent that at least three nuclear body proteins are upregulated by interferons (KORIOTH et al. 1995; CHELBI-ALIX et al. 1995); more detailed investigation of PML has revealed that at least in this case the phenomenon is mediated through specific response elements contained within the promoter region (STADLER et al. 1995).

Recently, the structure of the RING motif of PML in solution has been solved by NMR. Zinc binding was confirmed and found to be essential for folding of the domain, which contains two zinc atoms held in a cross-brace pattern, stabilised by a hydrophobic core region (BORDEN et al. 1995). This analysis has permitted very specific site-directed mutagenesis studies exploring the function of the PML RING finger and its role in nuclear body formation (BORDEN et al. 1995). Mutations of cysteines involved in either the first or second zinc binding site led to inability to form PML nuclear bodies (KASTNER et al. 1992; BORDEN et al. 1995), whereas mutations involving surface amino acids far from the zinc binding sites had no such effect (BORDEN et al. 1995). These studies lend weight to the view that the RING finger domain is important for protein-protein interactions which may be essential for the integrity of the nuclear bodies. This interaction may involve other PML molecules or other components of the nuclear domain. Electrostatic surface potential calculations indicate that the PML RING domain is nearly uniformly positive and hence a surface charge component may also be important for nuclear body formation (BORDEN et al. 1995). Studies involving mutations of the coiled-coil domain of PML or PML-RARα have implicated this region in PML-RARα/PML heterodimer and PML-RARα homodimer formation (KASTNER et al. 1992; PEREZ et al. 1993). More recent work suggests that the coiled-coil may also represent an important interface for wild-type PML homodimerisation (N. BODDY, personal communication; BORDEN et al. 1996), possibly leaving the RING finger and B-boxes free to interact with other motifs within the nuclear body. In support of this view, point mutations in either B-box domain also prevent PML nuclear body formation, without actually disrupting dimerisation with wild-type PML (BORDEN et al. 1996). It is becoming clear that understanding the role of the nuclear bodies, their composition, the determinants of their stability and the relationship with PML may be critical to understanding the pathogenesis of APL and its response to ATRA.

5 The RARα Protein and Retinoid Signalling Pathways

In order to understand the significance of *RARα* disruption in APL it is important to consider the normal mechanisms mediating retinoid signalling within the cell and their functional role. For many years it has been appreciated that retinoids exert profound effects on morphogenesis and differentiation which suggest they play an important role in embryogenesis (BROCKES 1990, for review and references therein). Furthermore, retinoids can achieve significant differentiating effects in several tumour cell types, including teratocarcinomas and myeloid leukaemias. Retinoid activity is mediated by two distinct families of nuclear receptors: the retinoic acid receptors (RARs : RARα, RARβ, RARγ), which can bind both ATRA and 9-*cis* RA, and the retinoid-X receptors (RXRs : RXRα, RXRβ, RXRγ), for which 9-*cis* RA is the only high affinity ligand (LEHMANN et al. 1992; LEID et al. 1992a). The retinoid receptors may exist as monomers, homodimers or RAR/RXR heterodimers (LEID et al. 1992a; AU-FLIEGNER et al. 1993), conferring their activity by binding to specific DNA response elements which leads to the activation or repression of target genes (BEATO 1989; LEID et al. 1992a; STUNNENBERG 1993 for reviews and references therein).

Steroid hormone receptors may be divided into two broad groups on the basis of their preferred dimerisation partner, DNA binding characteristics and features of the DNA binding motif (BEATO 1989; STUNNENBERG 1993). Hormone receptor response elements are composed of a hexanucleotide motif that may be repeated in a direct, inverted or palindromic pattern, separated by a variable number of spacer nucleotides (BEATO 1989; LEID et al. 1992a; STUNNENBERG 1993). Analysis of naturally occurring and synthetic response elements linked to reporter genes suggest that type I receptors, such as glucocorticoid, progesterone and oestrogen receptors, bind preferentially to palindromic motifs separated by three nucleotides. Hence such receptors are believed to bind DNA as homodimers in a "head to head" fashion with dimerisation occurring at the DNA binding domains (STUNNENBERG 1993). In contrast, type II receptors, such as those for thyroid hormone (TR), vitamin D (VDR) and the RARs, require the presence of RXR for high affinity DNA binding (ZHANG et al. 1992; KLIEWER et al. 1992; LEID et al. 1992a,b; STUNNENBERG 1993). RXR/type II receptor heterodimers bind to response elements which form a direct repeat pattern, favouring a "head to tail" binding model (STUNNENBERG 1993). The specificity of the DNA interaction is conferred by differences in the sequence and orientation of the hexanucleotide motif and also by variation in the number of spacer nucleotides (UMESONO et al. 1991; LEID et al. 1992a; STUNNENBERG 1993). Hence, VDR/RXR binds preferentially to a DR+3 motif (direct repeat with three spacer nucleotides), TR/RXR to a DR+4 element and RAR/RXR to DR+1, DR+2 or DR+5 motifs. In contrast RXR homodimers bind preferentially to a DR+1 element (UMESONO et al. 1991; STUNNENBERG 1993; KUROKAWA et al. 1994). Although RXR binds to a range of response elements in partnership with RAR, recent studies suggest that RXR ligand responses are not mediated through such a complex; indeed, the formation of RAR/RXR heterodimers renders the interaction of RXR with its specific ligands unfavourable (KUROKAWA et al. 1994; FORMAN et al. 1995). RXR responses may, therefore, be mediated through RXR homodimers acting at DR+1 response elements (KUROKAWA et al. 1994; FORMAN et al. 1995) and possibly through RXR/Nurr-1 or RXR/NGFI-B heterodimers at NGFI-B response elements (FORMAN et al. 1995). Ligand-bound RXR homodimers lead to transcriptional activation at DR+1 sites, in contrast to RAR/RXR heterodimers that bind such elements with greater affinity but have a repressive effect (KUROKAWA et al. 1994). Recent studies have demonstrated that the orientation of RAR/RXR heterodimers at DNA binding sites is a critical determinant of response, influencing release of the corepresser N-COR from the RAR hinge region in the presence of ligand (Kurokawa et al. 1994, 1995; Hörlein et al. 1995). At DR + 1 elements RAR binds the motif upstream of RXR, associated with transcriptional repression with persistent N-COR binding despite the presence of ligand to RAR. Whereas at DR + 5 elements, RXR is the upstream partner and RAR ligand binding leads to N-COR release and transcriptional activation.

Review of the mechanisms of retinoid induced signalling reveals multiplicity at all levels of the pathway which may be essential to achieve their wide ranging and far-reaching effects (Leid et al. 1992a). First, there is variability in ligand preferences for activation of a particular retinoid pathway. Both ATRA and 9-*cis* RA can activate RARs, whereas only 9-*cis* RA can bind to RXRs (Lehmann et al. 1992). It can be envisaged that a relative excess of a particular retinoid might lead to changes in response element activation patterns. For example, a rise in 9-*cis* RA in the cell might favour increased RXR homodimer activation at DR + 1 elements, possibly triggering a different repertoire of responses than if RAR/RXR-mediated responses had been favoured (Leid et al. 1992a; Schrader et al. 1993). Levels of retinoids within the cell may be modulated by binding proteins, whose activity in the adult and during embryogenesis may be temporally and spatially restricted (Leid et al. 1992a). Cellular retinol binding proteins (CRBPs) bind preferentially to retinol; whereas cellular retinoic acid binding proteins (CRABPs) bind ATRA rather than 9-*cis* RA (Leid et al. 1992a). The CRABPs contain a retinoic acid inducible promoter region and have been considered to contribute to resistance to long-term ATRA therapy in APL (Warrell 1993).

The RARs and RXRs exist in a series of subtypes which share significant sequence homology across species, but show little similarity between subtypes within the same species (Leid et al. 1992a). They vary in their pattern of distribution: for example, RARα is ubiquitously expressed, RARβ is present in a variety of epithelial cell types, whereas expression of RARγ appears to be largely confined to the skin (Collins et al. 1990). These spatial differences may be important for pattern formation during embryogenesis (Leid et al. 1992a). Further variability amongst the retinoid receptor subtypes may be generated as a result of two alternative promoter sites, which vary in retinoic acid inducibility, as well as by using alternative splicing amongst the 5′ exons. This creates a series of isoforms with variable NH_2-terminal A domains fused to common B to F regions (Leroy et al. 1991). This is particularly interesting since the rearrangement in APL also preserves the B to F domain of RARα, substituting the A domain for PML sequence. The A domain of the RARs may confer tissue specificity and ligand independent trans-activation function (Leroy et al. 1991; Leid et al. 1992a).

Therefore, in summary, the diversity demonstrated at all levels of the retinoid pathway is believed to facilitate the wide range of influences retinoids have on development (Leroy et al. 1991; Leid et al. 1992a). A series of RAR isoforms that vary in their tissue distribution and possibly in their developmental stage of expression can potentially heterodimerise with a range of variably expressed RXR isoforms to generate a considerable repertoire of functional responses (Leid et al. 1992a). Heterogeneity within the system is further enhanced by the potential diversity of the response elements created by the sequence and orientation of the recognition motif with variable spacing elements. The binding affinities and preferences of various RAR/RXR isoform heterodimers might be expected to differ across a range of potential retinoid response elements (LEID et al. 1992a; STUNNENBERG 1993). The character of the response may also be influenced by the relative activity of co-activator and co-repressor proteins that bind nuclear receptors in a ligand- and polarity-dependent fashion (LE DOUARIN et al. 1995; KUROKAWA et al. 1995; HÖRLEIN et al. 1995; DON CHEN and EVANS 1995); by the nature of the isoform A domain and further modulated by the relative ambient levels of ATRA, 9-*cis* RA or other retinoids (LEROY et al. 1991; LEID et al. 1992a; SCHRADER et al. 1993). Further diversity may be created by the tendency for RARs to heterodimerise with TRs and competition for limiting amounts of RXR for optimal DNA binding (BARRETINO et al. 1993).

Characterisation of the retinoid receptors has permitted more detailed investigation to determine whether they possess a physiological role in cellular function and tissue development. Knockout experiments in mice suggest that integrity of the RXRα pathway is fundamental to normal embryonic development. A homozygous defect proved lethal in utero due to cardiac malformation, mimicking the phenotype of vitamin A deficiency (KASTNER et al. 1994; SUCOV et al. 1994). Similar experiments involving RARs have led to more subtle defects; therefore the severe manifestations associated with loss of RXRα function could partly reflect disruption of VDR, TR and possibly Nurr 1/NGFI-B-dependent pathways in addition to any effect on retinoid response elements. Experiments achieving targeted expression of mutant RARs, associated with dominant negative activity, have demonstrated that integrity of the retinoid pathways is essential for normal embryonic skin development (IMAKADO et al. 1995; SAITOU et al. 1995). It is likely that this approach will further clarify and define the role of retinoid pathways not only in normal embryonic development but also in the postnatal period (ANDERSEN and ROSENFELD 1995). For example, transfection of COOH-terminally truncated RARα into mouse bone marrow cells or the FDCP haematopoietic pluripotential cell line inhibits the activity of wild-type receptor and leads to maturation arrest at the promyelocyte stage (TSAI and COLLINS 1993). It is this phenomenon that suggests that RARα is essential for normal myeloid differentiation and that disruption of the retinoid pathways plays a critical part in the pathogenesis and subsequent phenotype of APL.

6 Variant Translocations in Acute Promyelocytic Leukaemia

In APL, t(15;17) is frequently the only cytogenetic abnormality detected (MITELMAN 1994; BERGER et al 1991). This has led to the suggestion that this reciprocal translocation with its associated *PML*/*RAR*α rearrangement is by itself sufficient to cause leukaemic transformation (GRIGNANI et al. 1993) particularly as concomitant *ras* or *p*53 mutations in APL are extremely rare (LONGO et al. 1993). This view initially appears to contrast with that considered to underlie other acute leukaemias, in which a multistep process is envisaged leading to defects in both growth and differentiation (Sawyers et al. 1991).

Some patients with APL have been found to exhibit a normal karyotype. In many cases this may represent a failure of cytogenetic technique, in which purely direct examination has been employed that only effectively analyses the karyotype of normal marrow elements. However, in some patients the karyotype may be normal despite prior culture of the cells; in these cases RT-PCR or FISH may detect the *PML/RARα* fusion suggesting the presence of a cryptic rearrangement below the resolution of conventional cytogenetic assessment (Grimwade et al. 1996b). Similarly, in occasional cases in which variant cytogenetic abnormalities have been detected with no apparent t(15;17), or situations in which chromosomes 15 and/or 17 are involved as part of a three- or four- way translocation, molecular studies have nevertheless demonstrated a *PML/RARα* rearrangement, consistent with the view that it is this mediating the disease (Chen et al. 1994b; McKinney et al. 1994; Borrow et al. 1994; Hiorns et al. 1994).

Recently, two rare variant translocations have been described in patients with morphological APL, in which *RARα* is disrupted but fused to a partner other *PML*. The t(11;17)(q23;q21), which is described in detail in another review in this volume, involves a novel zinc finger gene *PLZF* (Chen et al. 1993a,b, 1994a). This disease, in contrast to that mediated by a *PML/RARα* rearrangement, appears to demonstrate a poor response to ATRA and is considered to confer an adverse prognosis, although the number of cases of t(11;17) APL described so far is small (Licht et al. 1995). More recently, a case of APL has been described involving a 5;17 translocation (q32;q21), which leads to the formation of a nucleophosmin-RARα fusion product (Redner et al. 1996). Nucleophosmin (NPM), a nucleolar phosphoprotein thought to be involved in RNA processing or packaging, has also been found to be fused to a tyrosine kinase catalytic domain of a novel gene, *ALK*, in Ki-1-positive anaplastic large cell lymphoma associated with t(2;5) (Morris et al. 1994; also described in detail in this volume). Further elucidation of the functional characteristics of these variant fusion proteins is likely to provide considerable insight into the pathogenesis of t(15;17) associated APL and the mechanisms mediating ATRA response. That no cases of APL with variant cytogenetics have been found with *PML*, *PLZF* or *NPM* fused to a novel gene suggest that disruption of these genes is not critical to the development of the promyelocytic phenotype, but may be essential for the process of leukaemic transformation. The involvement of *RARα* in 15;17 and the 5;17 and 11;17 cytogenetic variants suggest that disruption of this gene may be a prerequisite for the differentiation block in myeloid development that characterises APL.

7 Mechanisms of Leukaemogenesis and the All-*trans* Retinoic Acid Response

The t(15;17) translocation is unique to APL and hence is considered critical to the development of this subtype of AML. The condition usually arises de novo, although occasional secondary cases have been reported, particularly following the use of cytotoxic agents that interact with topoisomerases (Bhavnani et al. 1994). This is interesting in view of the topoisomerase I and II sites in the breakpoint regions (Dong et al. 1993; Tashiro et al. 1994). PML-RARα is the only fusion product resulting from the translocation that is invariably transcribed and it retains the potential important functional domains of both PML and RARα; thus, it is the most likely mediator of leukaemic transformation. Initially, PML was considered to be a transcription factor in view of its zinc finger domains, hence APL was thought to result from disruption of PML- and/or RARα- dependent transcription pathways. The recent characterisation of the rare APL cytogenetic variants t(5;17) and t(11;17) demonstrates that disruption of RARα is a prerequisite for the disease. Furthermore, in both t(5;17) and t(11;17) variant translocations, the breakpoint lies within the second intron of RARα as in the t(15;17) (Redner et al. 1996; Chen et al. 1993a,b), consistent with the view that this region represents a recombination hot-spot (Tashiro et al. 1994). It also suggests that recombination at this breakpoint site, which results in retention of the DNA binding, heterodimerisation and ligand binding regions of the receptor, is essential for the functional activity of the fusion protein. The involvement of *RARα*, even in APLs with variant cytogenetics, suggest that disruption of this gene is critical to the development of the promyelocytic leukaemia phenotype and implicates RARα in the normal process of myeloid differentiation.

For many years it has been appreciated that pharmacological doses of retinoids can cause differentiation of myeloid leukaemic blasts, particularly those with M2/M3 morphology (Breitman et al. 1981; Huang et al. 1988; Castaigne et al. 1990; Chomienne et al. 1990; Sakashita et al. 1993). Recently, an HL60 cell line (HL60R) that is relatively resistant to the differentiating effects of retinoids, in comparison to wild-type HL60, was found to harbour a COOH-terminally mutated RARα (Robertson et al. 1992a). This leads to a reduction in ligand binding affinity, but probably has little effect on the ability to form RXR heterodimers. It was found that the differentiation response of HL60R to ATRA could be restored by transfection of wild-type RARs or RXRα (Robertson et al. 1992b). This indicates that integrity of the retinoid receptor pathways, of which RARα and RXRβ are probably the major components in haematopoietic cells (Robertson et al. 1992b; Sakashita et al. 1993), is essential for mediating the response to ATRA in HL60 cells and suggests that physiological levels of retinoic acid might be important to sustain normal myeloid differentiation. In favour of such a hypothesis, it was found that transfection of dominant negative mutant *RARα* , with an interrupted ligand-binding domain, into murine pluripotential factor-dependent haematopoietic stem cells (FDCP) or into mouse bone marrow led to a block in differentiation at the promyelocyte stage which could be overcome by pharmacological doses of ATRA (TSAI and COLLINS 1993). This ATRA response was presumably mediated by residual normal RAR/RXR heterodimers or RAR/RAR homodimers. Similarly, transfection of *PML-RARα* into the U937 promonocytic cell line prevented vitamin D_3-induced monocytic differentiation, a block which could also be overcome by addition of ATRA (GRIGNANI et al. 1993; TESTA et al. 1994). Expression of *PML-RARα* in U937 cells serves as a model, appearing to reproduce features of APL: the chimaeric gene leads to increased proliferation associated with a reduced rate of apoptosis and introduces a block in differentiation that can be overcome by ATRA. This model provides further, indirect evidence that *PML-RARα* is the leukaemogenic transcript.

It has been assumed that PML-RARα mediates leukaemogenesis by exerting "dominant negative" activity on PML- and/or RARα-dependent pathways. The aforementioned studies of ATRA-resistant HL60R cells (ROBERTSON et al. 1992b) and transfection studies in FDCP cells or mouse bone marrow (TSAI and COLLINS, 1993) demonstrate that aberrant RARα receptors can lead to such "dominant negative" activity. However, in all of these systems the receptor defect lies within the ligand binding site, which therefore does not fully mimic the situation in APL in which ligand binding and RXR heterodimerisation regions of RARα remain intact. A number of groups have attempted to address whether PML-RARα can mediate a "dominant negative" effect on retinoid signalling using transient transfection studies. The experiments have not proved particularly enlightening as to the processes underlying leukaemogenesis in APL or the mechanisms mediating ATRA-induced differentiation since the results were highly dependent upon the cell type and reporter construct/retinoid response element employed. This is not unexpected since heterogeneity within the retinoid signalling pathway is believed to confer functional diversity within the system (LEID et al. 1992a). DE THÉ et al. (1991) found that PML-RARα is a relatively poor trans-activator at physiological levels of ATRA, suggesting that it could block normal retinoic acid-dependent myeloid differentiation, which could be overcome by supraphysiological levels of ATRA. The results of PANDOLFI et al. (1991) also implicated PML-RARα as a repressor of normal retinoid signalling in the absence of ligand; but in these experiments the presence of ATRA led to more potent trans-activation of retinoid response elements by the fusion protein than wild-type receptor. Again these experiments suggest that the differential activity of PML-RARα in the presence or absence of ligand may account for the cause of the leukaemia and its response to ATRA. KAKIZUKA et al. (1991) found that in some cell types, e.g. CV1, the fusion protein led to activation at retinoid response elements and that exposure to ATRA led to further enhancement of this phenomenon in comparison to wild-type receptor activation. By contrast, in myeloid cell lines such as HL60, PML-RARα had a repressive effect in the absence of ligand and caused activation upon treatment with ATRA, consistent with the findings of DE THÉ (1991). Using an NH_2-truncated RARα construct lacking the A domain, KASTNER et al. (1992) demonstrated that the trans-activation properties of the fusion protein are modulated by the presence of PML and do not merely reflect an effect of NH_2-terminal truncation.

Although no consistent message emerges from this series of experiments they do demonstrate a number of points. PML-RARα is clearly capable of activating or repressing various retinoid response elements and has both promoter and cell type specificities. The fusion product is capable of binding retinoic acid which may alter its response at retinoid response elements, suggesting that a differential effect of PML-RARα at physiological and pharmacological levels of ATRA might explain the differentiation block at low levels of ligand, with induction of differentiation at therapeutic doses. Furthermore, the experiments of KASTNER et al. (1992) suggest that APL is not solely the result of NH_2-terminal truncation of RARα. The presence of PML in the fusion protein clearly influences its behaviour at retinoid response elements and more recent data suggest that PML may play an important role in leukaemic transformation (MU et al. 1994; LIU et al. 1995; KOKEN et al. 1995).

With the development of PML antisera, which demonstrated inclusion of wild-type PML within nuclear bodies, it became apparent that in APL there is significant disruption of nuclear architecture (KASTNER et al. 1992; DANIEL et al. 1993; DYCK et al. 1994; WEIS et al. 1994; KOKEN et al. 1994). In normal cells there are approximately 10–30 nuclear domains, whereas in APL cells the majority of domains are disrupted, leading to a microparticulate pattern of PML within the nucleus (DYCK et al. 1994; WEIS et al. 1994; KOKEN et al. 1994). Electron microscopy studies demonstrate the presence of some residual 0.3-0.5 µm nuclear domains lying adjacent to chromatin; however these are mostly replaced by large numbers of small electron dense particles not greater than 0.1 µm which are tightly bound to chromatin (WEIS et al. 1994). These smaller structures appear to correspond to the microparticulate pattern of PML and some at least contain other nuclear domain constituents such as SP100 (WEIS et al. 1994). Immunofluorescence studies have demonstrated that PML-RARα also colocalises to the disrupted nuclear bodies (WEIS et al. 1994), reflecting its ability to heterodimerise with PML (DYCK et al. 1994). Transfection studies suggest that PML-RARα is also capable of forming heterodimers with RXR (PEREZ et al. 1993). This interaction has been confirmed in vivo and may account for the sequestration of RXR within the nuclear domains noted in immunofluorescence studies performed in APL cells (WEIS et al. 1994). With subsequent treatment with ATRA, RXR becomes dissociated from the abnormal APL-associated nuclear bodies and nuclear structure returns to normal with the reconstitution of 10–30 nuclear domains (WEIS et al. 1994). This effect is not the result of a significant change in the levels of PML, RARα or PML-RARα induced by ATRA (WEIS et al. 1994; SARKAR et al. 1994).

This sequence of events suggests a further model, based on the sequestration of RXR, that can explain both the block in myeloid differentiation associated with APL and the response to ATRA therapy. Sequestration of RXR by PML-RARα into the abnormal nuclear bodies may have two potential effects. First, less RXR may be available to mediate the high-affinity DNA binding, not only of residual wild-type RARα but also of VDR and TR to their respective response elements. Their activity may be critical to achieve normal myeloid differentiation and account for the block at the promyelocyte stage associated with APL. RARα, VDR and TR all seem to play a role in myeloid differentiation and may be important in modulating its pattern. Integrity of RARα function may permit progression along the granulocytic pathway (Tsai and Collins 1993), vitamin D-mediated responses appear to drive differentiation down the monocytic pathway (Grignani et al. 1993; Testa et al. 1994), whereas the integrity of TR activity may be important for erythroid differentiation (Damm et al. 1989). Second, as PML-RARα/RXR heterodimers are capable of binding to retinoid response elements (Perez et al. 1993), it is possible to envisage that the binding specificities and *trans*-activation properties of the fusion product might differ from those of wild-type RARα/RXR and hence abnormal patterns of activation and/or repression of retinoid response pathways might contribute to the leukaemic phenotype, as was suggested by various transient transfection studies (Pandolfi et al. 1991; de Thé et al. 1991; Kakizuka et al. 1991; Kastner et al. 1992). Treatment of APL cells with ATRA might lead to a conformational change in PML-RARα that enables reformation of nuclear bodies and impairs PML-RARα/RXR heterodimer formation such that RXR is released. Free RXR might then facilitate differentiation by interacting with wild-type RARα, VDR and TR. It is also possible that part of the ATRA effect occurs independently of RXR, mediated by RARα homodimers.

This model therefore proposes sequestration of RXR as the major factor underlying the differentiation block in APL with its release on ATRA therapy then mediating terminal myeloid differentiation. In support of such a model are the results of experiments whereby U937 cells were transfected with *PML-RARα* (Grignani et al. 1993; Testa et al. 1994). This cell line can normally be induced to differentiate into more mature monocytic cells in the presence of vitamin D_3, but this response is blocked on expression of the fusion protein (Grignani et al. 1993; Testa et al. 1994). Since VDR requires RXR for optimal activity (Kliewer et al. 1992), this failure in differentiation may also be explained by sequestration of RXR in the transfected cells. On treatment with ATRA, monocytic differentiation to D_3 exposure is restored (Testa et al. 1994), presumably secondary to release of sequestered RXR.

It has also been suggested that disruption of the PML-containing nuclear bodies may be the primary phenomenon in APL pathogenesis and that failure to reconstitute the nuclear domains is associated with resistance to ATRA and hence failure to achieve differentiation (Dyck et al. 1994). In this context it is possible to envisage that the abnormal localisation of one or more factors associated with the nuclear domains might lead to leukaemogenesis; this could involve PML, RXR or another as yet uncharacterised nuclear domain constituent. The disruption of nuclear domains in APL is likely to be mediated by the fusion protein and possibly aberrant PML molecules which have variable COOH-terminal truncations involving coiled-coil and B-box domains. Mutational analyses suggest that the integrity of RING, B-box and coiled-coil domains of wild-type PML is important for the stability of the nuclear domains (Kastner et al. 1992; Borden et al. 1995,1996). The coiled-coil is known to be important for PML/PML-RARα heterodimer and PML-RARα homodimer formation (Kastner et al. 1992; Perez et al. 1993). Recent studies suggest that it is also required for PML homodimerisation (N. Boddy, unpublished), leaving RING and B-box domains free to interact with other PML molecules or other constituents of the nuclear domain. Electrostatic charge is also believed to be important for the stability of the nuclear bodies (Borden et al. 1995); therefore, the differences in structure and/or charge characteristics of the fusion proteins might account for their disruption in APL. Treatment with ATRA with consequent ligand binding might induce a further conformational change in PML-RARα, rendering the interaction with RXR less favourable and permitting reconstitution of normal nuclear domains.

Such a scheme suggests that disruption of the nuclear bodies is not the primary abnormality mediating leukaemogenesis, but merely a secondary phenomenon reflecting an association between PML-RARα and PML within the nuclear domains. Whether this view is valid will become apparent with further characterisation of the role of PLZF and NPM in variant translocations in APL and whether these conditions are associated with disruption of PML nuclear bodies. There is already early evidence to suggest that PLZF, although localised within the nucleus, is not associated with the nuclear bodies (Licht et al. 1995).

The normal role of PML, apart from representing a constituent of the nuclear domains, is unclear. However, recently PML has been found to demonstrate anti-oncogenic activity in a variety of transformation assays (Mu et al. 1994). Transfection of *PML* into APL-derived NB4 cells suppressed their anchorage-independent growth on soft agar and tumourigenicity in nude mice. PML was found to suppress transformation of rat embryo fibroblasts by Ha-*ras* in the presence of mutant *p*53 or c-*myc*. Furthermore PML prevented the transformation of NIH 3T3 cells by activated *neu* oncogene, an effect that was abrogated by cotransfection with *PML-RARα*. Whether this activity is directly attributable to PML or is a secondary phenomenon due to PML aggregates functioning in a manner analogous to the nuclear bodies is unclear. The coiled-coil domain of PML shares homology with the Jun/Fos dimerisation interface and could form a potential site of interaction with transcription factors that could be involved in mediating this growth-suppressive activity. That PML plays an active role in leukaemogenesis is suggested by the involvement of other RING finger proteins such as RFP and TIFI in oncogenic fusion products.

Leukaemia is considered to develop as a result of a multistep process combining defects of growth and differentiation (Sawyers et al. 1991). In APL t(15;17) is often the sole cytogenetic abnormality (Mitelman 1994; Berger et al 1991) and associated *p*53 or *ras* mutations are rare (Longo et al 1993). Since PML appears to demonstrate growth suppressor activity, the *PML/RARα* rearrangement may be sufficient by itself to mediate leukaemogenesis (Mu et al. 1994). Disruption of *PML* could lead to cell transformation associated with reduced apoptosis and concomitant interruption of RARα responsive pathways and sequestration of RXR cause the characteristic differentiation block at the promyelocyte stage (Fig. 3). This differentiation block may be enforced by inhibition of AP-1 activity by PML-RARα (Doucas et al. 1993)and high annexin VIII levels that inhibit phospholipase A (Sarkar et al. 1994); both of these factors are subsequently reversed on treatment with ATRA. Whether PML plays an active or merely permissive role in the

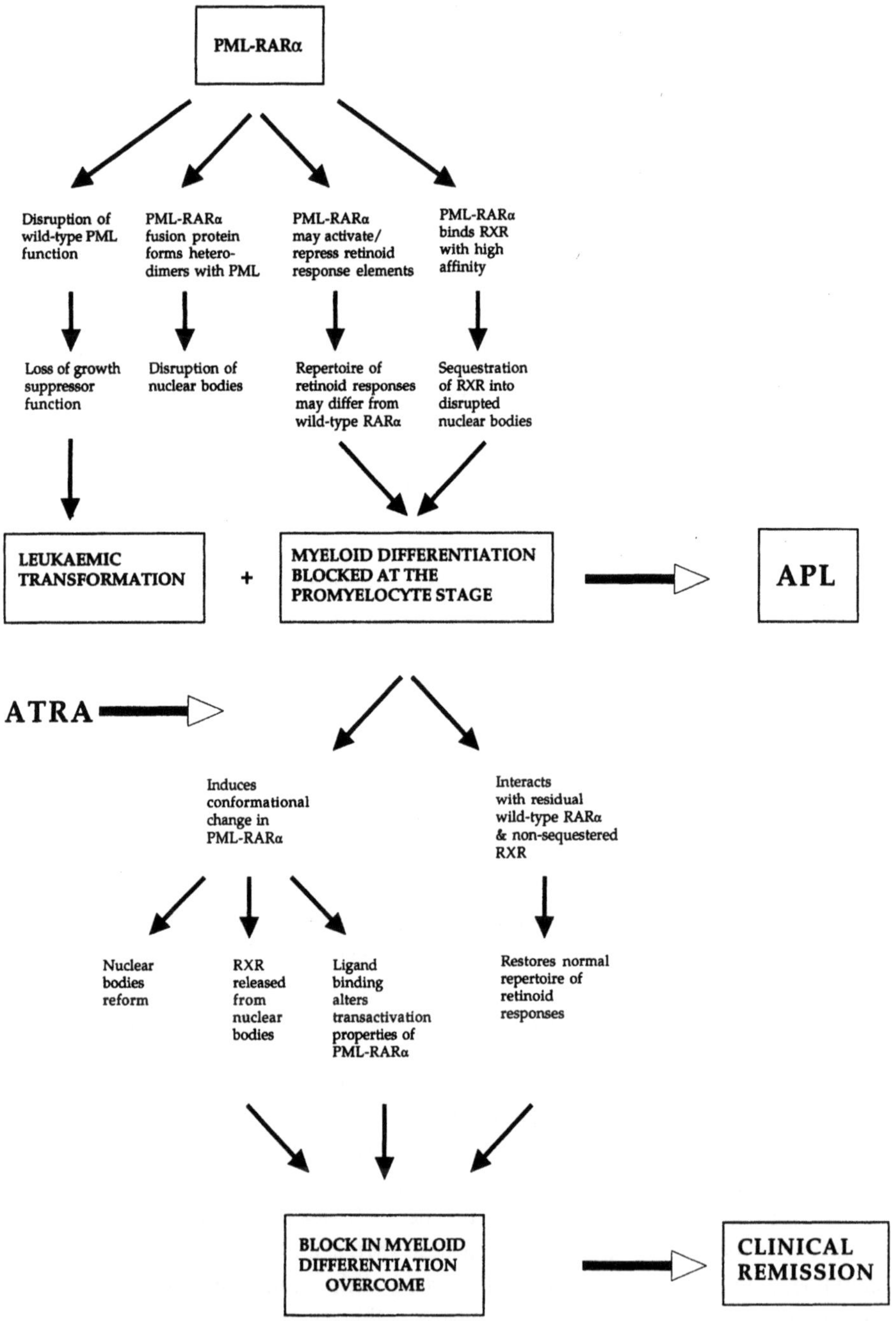

Fig. 3. Model for the pathogenesis of acute promyelocytic leukemia (APL) and the response to retinoids (ATRA)

development of APL will become apparent with further characterisation of the structure and function of the nuclear domains, establishment of the normal role of PLZF and NPM and determination of their structural and functional similarities to PML.

8 The Role of Molecular Techniques in the Management of Patients with Acute Promyelocytic Leukaemia

Molecular studies may prove a useful adjunct to the care of patients with APL. Demonstration of the presence of a *PML/RARα* rearrangement, by techniques such as RT-PCR, is particularly valuable in patients in whom cytogenetics at diagnosis has failed or revealed a variant or normal karyotype. This serves to confirm the diagnosis of APL at the molecular level and identifies a group with a favourable differentiation response to ATRA (Miller et al. 1992).

Recent studies have suggested that characterization of the *PML* breakpoint pattern may also provide independent prognostic information. A subgroup of APL with *PML* exon 6 (*bcr*2) breakpoints has been found to exhibit reduced sensitivity to ATRA, at least in vitro; whether this is of any clinical significance remains to be determined in a larger number of cases (Gallagher et al. 1995). Furthermore, two groups have suggested that the 5′ (*bcr*3) pattern is associated with a worse prognosis, although these studies involved patients that had received rather heterogeneous treatment approaches (Borrow et al. 1992; Huang et al. 1993). More recently, the *bcr*3 pattern was found to predict an increased risk of relapse in a group of patients treated more uniformly with ATRA and chemotherapy (Vahdat et al. 1994). Should this phenomenon be confirmed in larger ongoing trials, it would suggest that all patients with APL should have the *PML* breakpoint pattern characterised at diagnosis. Furthermore, modifications to the treatment protocol of patients with a *bcr*3 breakpoint should be considered, such as transplantation in first complete remission (CR), which is not currently recommended for the treatment of APL.

The sensitivity of nested RT-PCR techniques also affords the opportunity to monitor minimal residual disease (MRD) during the treatment period and after completion of consolidation therapy. Early studies of MRD suggested that persistent PCR positivity 2–4 months after achievement of CR predicted relapse; whereas PCR negativity was associated with a good prognosis (Lo Coco et al. 1992). Accordingly patients with APL in long-term remission were found to be PCR negative (Diverio et al. 1993). However, a number of early studies of MRD in APL contained significant numbers of patients treated in relapse or in whom ATRA was the sole therapy. Such a group would be expected to demonstrate an increased rate of persistent PCR positivity associated with a high risk of relapse. It is now clear that combined therapy with ATRA and chemotherapy represents optimal management of APL at present (Fenaux et al. 1994) and is associated with the achievement of PCR negativity in the vast majority of patients (Miller et al. 1993). Therefore the conclusions relating to the predictive power of PCR results alluded to in the early studies can not be directly transposed to apply to current treatment protocols. Indeed it is now clear that a significant number of ATRA and chemotherapy treated patients may relapse despite a series of PCR negative bone marrow examinations in remission (Grimwade et al. 1996a). This highlights the relative insensitivity of RT-PCR assays currently employed in APL, compared to those for diseases such as chronic granulocytic leukaemia. This may reflect the fragility of APL cells containing a relatively high concentration of inherent RNase activity, but is probably largely due to inefficiency of the reverse transcription step (Seale et al. 1994). Attempts to increase PCR sensitivity in APL have only succeeded in identifying PCR positivity in patients in long-term remission (Tobal et al. 1995). This is interesting from the point of view of the biology of leukaemia remission and agrees with the presence of detectable *Runt-MTG8* transcripts in t(8;21) AML M2 patients in long-term remission (Nucifora et al. 1993); however, such information is of little value in planning the care of individual patients, since the prognostic significance of such low-level PCR positivity is unclear. In fact it is the relative insensitivity of generally employed RT-PCR assays, which can detect one APL cell in 10^4–10^5 normal cells, that may actually prove to be advantageous, enabling finer tailoring of care to individual patients. At the sensitivity level of current assays it is clear that, in patients treated with chemotherapy and ATRA, persistent PCR positivity on completing consolidation therapy predicts relapse (Miller et al. 1993). The optimal treatment for such patients is currently being addressed by the Italian AIDA study (Avvisati et al. 1994). Recurrence of PCR positivity for *PML-RARα* transcripts whilst in remission has been found to herald relapse within 1 year, although in some cases relapse occurs within only a few weeks (Korninger et al. 1994; Grimwade et al. 1996a). There is currently no evidence that reinstituting therapy at the time of recurrence of PCR positivity confers any survival advantage, compared to treatment commenced at the time of frank relapse. Therefore there may be little merit in performing frequent marrow aspirates in remission for PCR assessment purposes unless the patient has an allogeneic transplant option. In view of the failure to detect MRD post consolidation in all patients who ultimately relapse, current trials are addressing whether the rapidity of clearance of PCR-detectable disease is a better indicator of long-term prognosis. Should this prove to be the case it may permit further rationalisation of treatment to individual patients in addition to that afforded by breakpoint characterisation. Ideally those patients with a significant risk of relapse may be selected for more aggressive therapy, whereas those considered to have a good prognosis might be spared excessive treatment associated with increased morbidity and expense.

9 Summary

The vast majority of cases of APL are associated with t(15;17) leading to the formation of PML-RARα, RARα-PML and aberrant PML fusion products. PML-RARα is invariably transcribed and is believed to mediate leukaemogenesis. PML was initially considered to be a transcription factor. However, characterisation of other RING finger containing proteins shows no direct evidence for DNA binding. The RING, B-box, and coiled-coil domains are more likely to represent sites of protein-protein interaction and may be critical for the stability of the multiprotein nuclear domains of which PML is an integral part. In APL the nuclear bodies become disrupted, presumably as a consequence of the presence of PML-RARα and aberrant PML proteins that might render the structure unstable. PML-RARα is capable of binding RXR and sequestering it into the disrupted nuclear domains. Sequestration of RXR would be expected to limit high affinity binding of VDR, TR and residual RARs to DNA response elements and might account for the block in myeloid differentiation at the promyelocyte stage that characterizes APL. Recently PML has been found to have growth suppressor/anti-oncogenic activity. It is unclear whether this is a property of PML itself or reflects a nonspecific function of the PML-associated nuclear domains. Hence the *PML/RARα* rearrangement alone may be sufficient to cause APL. Abnormal PML function may prevent its growth-suppressor activity, leading to leukaemic transformation; concomitant disruption of retinoid pathways due to sequestration of RXR and/or an abnormal repertoire and character of response element activation mediated by the fusion protein, causing the block in myeloid differentiation (Fig. 3). Disruption of RARα would be expected to account for the similar leukaemic phenotype associated with the t(5;17) and t(11;17) APL cytogenetic variants. Further characterisation of NPM and PLZF at the structural and functional level will determine whether PML and other proteins disrupted in APL associated translocations play an active or purely permissive role in leukaemogenesis and will help dissect the events leading to transformation from those causing blockade of myeloid differentiation and mediating the response to ATRA.

Acknowledgements. We would like to thank Beatrice Griffiths, Lynne Davies, Malcolm Parker and Paul Freemont for critical reading of the manuscript. Work on APL in our laboratory is supported by EEC grant BIOMED BMHI-CT92-0755 and the ICRF. David Grimwade is funded by the Medical Research Council.

References

Alcalay M, Zangrilli D, Pandolfi PP, Longo L, Mencarelli A, Giacomucci A, Rocchi M, Biondi A, Rambaldi A, Lo Coco F, Diverio D, Donti E, Grignani F, Pelicci PG (1991) Translocation breakpoint of acute promyelocytic leukemia lies within the retinoic acid receptor α locus. Proc Natl Acad Sci USA 88: 1977–1981

Alcalay M, Zangrilli D, Fagioli M, Pandolfi PP, Mencarelli A, Lo Coco F, Biondi A, Grignani F, Pelicci PG (1992) Expression pattern of the RARα-PML fusion gene in acute promyelocytic leukemia. Proc Natl Acad Sci USA 89: 4840–4844

Andersen B, Rosenfeld MG (1995) New wrinkles in retinoids. Nature 374: 118–119

Au-Fliegner M, Helmer E, Casanova J, Raaka BM, Samuels HH (1993) The conserved ninth C-terminal heptad in thyroid hormone and retinoic acid receptors mediates diverse responses by affecting heterodimer but not homodimer formation. Mol Cell Biol 13: 5725–5737

Avvisati G, Ten Cate JW, Mandelli F (1992) Acute promyelocytic leukaemia. Br J Haematol 81: 315–320

Avvisati G, Baccarani M, Ferrara F, Lazzarino M, Resegotti L, Mandelli F (1994) AIDA protocol (all-*trans* retinoic acid + idarubicin) in the treatment of newly diagnosed acute promyelocytic leukaemia: a pilot study of the Italian cooperative group GIMEMA. Blood 84 (Suppl 1): 380a

Bain BJ (1990) Leukaemia diagnosis. A guide to the FAB classification. Gower Medical, London, p 14

Barettino D, Bugge TH, Bartunek P, Vivanco Ruiz MdM, Sonntag-Buck V, Beug H, Zenke M, Stunnenberg HG (1993) Unliganded T_3R, but not its oncogenic variant v-erbA, suppresses RAR dependent transactivation by titrating out RXR. EMBO J 12: 1343–1354

Beato M (1989) Gene regulation by steroid hormones. Cell 56: 335–344

Bennett J, Catovsky D, Daniel MT, Flandrin G, Galton DAG, Gralnick NR, Sultan C (1976) [French-American-British (FAB) cooperative group]. Proposals for the classification of the acute leukaemias. Br J Haematol 33: 451–458

Berger R, Le Coniat M, Derré J, Vecchione D, Jonveaux P (1991) Cytogenetic studies in acute promyelocytic leukemia: a survey of secondary chromosomal abnormalities. Genes Chromosom Cancer 3: 332–337

Bhavnani M, Al Azzawi S, Liu Yin JA, Lucas GS (1994) Therapy-related acute promyelocytic leukaemia. Br J Haematol 86: 231–232

Biondi A, Rambaldi A, Pandolfi PP, Rossi V, Giudici G, Alcalay M, Lo Coco F, Diverio D, Pogliani EM, Lanzi EM, Mandelli F, Masera G, Barbui T, Pelicci PG (1992) Molecular monitoring of the myl/retinoic acid receptor-α fusion gene in acute promyelocytic leukemia by polymerase chain reaction. Blood 80: 492–497

Biondi A, Luciano A, Bassan R, Mininni D, Specchia G, Lanzi E, Castagna S, Cantu-Rajnoldi A, Liso V, Masera G, Barbui T, Rambaldi A (1993) CD2 expression correlates with microgranular acute promyelocytic leukemia (M3V) and not with PML gene breakpoint. Blood 82: 113a

Biondi A, Rovelli A, Cantu-Rajnoldi A, Fenu S, Basso G, Luciano A, Rondelli R, Mandelli F, Masera G, Testi AM (1994) Acute promyelocytic leukemia in children : experience of the Italian pediatric hematology and oncology group (AIEOP). Leukemia 8 (Suppl) : s66-s70

Biondi A, Luciano A, Bassan R, Mininni D, Specchia G, Lanzi E, Castagna S, Cantu-Rajnoldi A, Liso V, Masera G, Barbui T, Rambaldi A (1995) CD2 expression in acute promyelocytic leukemia is associated with microgranular morphology (FAB M3v) but not with any PML breakpoint. Leukemia 9: 1461–1466

Boddy MN, Howe K, Etkin LD, Solomon E, Freemont PS (1996) PIC 1, a novel ubiguitin-like protein which interacts with the PML component of a multiprotein complex that is disrupted in acute promyelocytic leukaemia. Oncogene (in press)

Borden KLB, Boddy MN, Lally J, O'Reilly NJ, Martin S, Howe K, Solomon E, Freemont PS (1995) The solution structure of the RING finger domain from the acute promyelocytic proto-oncoprotein PML. EMBO J 14: 1532–1541

Borden KLB, Lally JM, Martin SR, O'Reilly NJ, Solomon E, Freemont PS (1996) In vivo and in vitro characterization of the B1 and B2 zinc-binding domains from the acute promyelocytic leukemia protooncoprotein PML. Proc Natl Acad Sci USA 93: 1601–1606

Borrow J, Goddard AD, Sheer D, Solomon E (1990) Molecular analysis of acute promyelocytic leukemia breakpoint cluster region on chromosome 17. Science 249: 1577–1580

Borrow J, Goddard AD, Gibbons B, Katz F, Swirsky D, Fioretos T, Dube I, Winfield DA, Kingston J, Hagemeijer A, Rees JKH, Lister TA, Solomon E (1992) Diagnosis of acute promyelocytic leukaemia by RT-PCR: detection of PML-RARA and RARA-PML fusion transcripts. Br J Haematol 82: 529–540

Borrow J, Solomon E (1992) Molecular analysis of the t(15;17) translocation in acute promyelocytic leukaemia. Baillieres Clin Haematol 5: 833–856

Borrow J, Shipley J, Howe K, Kiely F, Goddard A, Sheer D, Srivastava A, Antony AC, Fioretos T, Mitelman F, Solomon E (1994) Molecular analysis of simple variant translocations in acute promyelocytic leukaemia. Genes Chromosom Cancer 9: 234–243

Brand NJ, Petkovich M, Chambon P (1990) Characterization of a functional promoter for the human retinoic acid receptor-alpha (hRAR-α). Nucleic Acids Res 18: 6799–6806

Breitman TR, Collins SJ, Keene BR (1981) Terminal differentiation of human promyelocytic leukemic cells in primary culture in response to retinoic acid. Blood 57: 1000–1004

Brockes J (1990) Reading the retinoid signals. Nature 345: 766–768

Burnett AK (1994) Karyotypically defined risk groups in acute myeloid leukaemia. Leuk Res 18: 889–890

Carvalho T, Seeler J-S, Ohman K, Jordan P, Pettersson U, Akusjärvi G, Carmo-Fonseca M, Dejean A (1995) Targeting of adenovirus E1A and E4-ORF3 proteins to nuclear matrix-associated PML bodies. J cell Biol 131: 45–56

Castaigne S, Chomienne C, Daniel MT, Ballerini P, Berger R, Fenaux P, Degos L (1990) All-trans retinoic acid as a differentiation therapy for acute promyelocytic leukemia. I. Clinical results. Blood 76: 1704–1709

Castoldi GL, Liso V, Specchia G, Tomasi P (1994) Acute promyelocytic leukemia : morphological aspects. Leukemia 8 (Suppl) : s27–s32

Chelbi-Alix MK, Pelicano L, Quignon F, Koken MHM, Venturini L, Stadler M, Pavlovic J, Degos L, and de Thé H (1995) Induction of the PML protein by interferons in normal and APL cells. Leukemia 9: 2027–2033

Chen S-J, Zelent A, Tong J-H, Yu H-Q, Wang Z-Y, Derré J, Berger R, Waxman S, Chen Z (1993a) Rearrangements of the retinoic acid receptor alpha and promyelocytic leukaemia zinc finger genes resulting from t(11;17) (q23;q21) in a patient with acute promyelocytic leukaemia. J Clin Invest 91: 2260–2267

Chen Z, Brand NJ, Chen A, Chen S-J, Tong J-H, Wang Z-Y, Waxman S, Zelent A (1993b) Fusion between a novel Kruppel-like zinc finger gene and the retinoic acid receptor-α locus due to a variant t(11;17) translocation associated with acute promyelocytic leukaemia. EMBO J 12: 1161–1167

Chen Z, Guidez F, Rousselot P, Agadir A, Chen S-J, Wang Z-Y, Degos L, Zelent A, Waxman S, Chomienne C (1994a) PLZF-RARα fusion proteins generated from the variant t(11;17) (q23;q21) translocation in acute promyelocytic leukemia inhibit ligand-dependent transactivation of wild-type retinoic acid receptors. Proc Natl Acad Sci USA 91: 1178–1182

Chen Z, Morgan R, Stone JF, Sandberg AA (1994b) Identification of complex t(15;17) in APL by FISH. Cancer Genet Cytogenet 72: 73–74

Chomienne C, Ballerini P, Balitrand N, Daniel MT, Fenaux P, Castaigne S, Degos L (1990) All-trans retinoic acid in acute promyelocytic leukemias. II. In vitro studies: structure-function relationship. Blood 76: 1710–1717

Claxton DF, Reading CL, Nagarajan L, Tsujimoto Y, Andersson BS, Estey E, Cork A, Huh YO, Trujillo J, Diesseroth AB (1992) Correlation of CD2 expression with PML gene breakpoints in patients with acute promyelocytic leukemia. Blood 80: 582–586

Collins SJ, Robertson KA, Mueller L (1990) Retinoic acid-induced granulocytic differentiation of HL-60 myeloid leukaemia cells is mediated directly through the retinoic acid receptor (RAR-α). Mol Cell Biol 10: 2154–2163

Damm K, Thompson CC, Evans RM (1989) Protein encoded by *v-erb A* functions as a thyroid hormone receptor antagonist. Nature 339: 593–597

Daniel MT, Koken M, Romagné O, Barbey S, Bazarbachi A, Stadler M, Guillemin MC, Degos L, Chomienne C, de Thé H (1993) PML protein expression in hematopoietic and acute promyelocytic leukemia cells. Blood 82: 1858–1867

Danielan PS, White R, Lees JA, Parker MG (1992) Identification of a conserved region required for hormone-dependent transcriptional activation by steroid hormone receptors. EMBO J 11: 1025–1033

de Thé H, Chomienne C, Lanotte M, Degos L, Dejean A (1990) The t(15;17) translocation of acute promyelocytic leukaemia fuses the retinoic acid receptor α gene to a novel transcribed locus. Nature 347: 558–561

de Thé H, Lavau C, Marchio A, Chomienne C, Degos L, Dejean A (1991) The PML-RARα fusion mRNA generated by the t(15;17) translocation in acute promyelocytic leukemia encodes a functionally altered RAR. Cell 66: 675–684

Di Noto R, Schiavone EM, Ferrara F, Manzo C, Lo Pardo C, Del Vecchio L (1994) Expression and ATRA-driven modulation of adhesion molecules in acute promyelocytic leukemia. Leukemia 8 (Suppl): s71-s76

Diverio D, Pandolfi PP, Biondi A, Avvisati G, Petti MC, Mandelli F, Pelicci PG, Lo Coco F (1993) Absence of reverse transcription-polymerase chain reaction detectable residual disease in patients with acute promyelocytic leukemia in long-term remission. Blood 82: 3556–3559

Don Chen J, Evans RM (1995) A transcriptional co-repressor that interacts with nuclear hormone receptors. Nature 377:454–457

Dong S, Geng J-P, Tong J-H, Wu Y, Cai J-R, Sun G-L, Chen S-R, Wang Z-Y, Larsen C-J, Berger R, Chen S-J, Chen Z (1993) Breakpoint clusters of the PML gene in acute promyelocytic leukemia : Primary structure of the reciprocal products of the PML-RARA gene in a patient with t(15;17). Genes, Chromosom Cancer 6: 133–139

Doucas V, Brockes JP, Yaniv M, de Thé H, Dejean A (1993) The PML-retinoic acid receptor α translocation converts the recptor from an inhibitor to a retinoic acid-dependent activator of transcription factor AP-1. Proc Natl Acad Sci USA 90: 9345–9349

Doucas V, Ishov AM, Romo A, Juguilon H, Weitzman MD, Evans RM, Maul GG (1996) Adenovirus replication is coupled with the dynamic properties of the PML nuclear structure. Genes Dev 10: 196–207

Dulic V, Riezman H (1989) Characterization of the *END 1* gene required for vacuole biogenesis and gluconeogenic growth of budding yeast. EMBO J 8: 1349–1359

Durand B, Saunders M, Gaudon C, Roy B, Losson R, Chambon P (1994) Activation function 2 (AF-2) of RAR and RXR: presence of a conserved autonomous constitutive activating domain and influence of the nature of the response element on AF-2 activity. EMBO J 13: 5370–5382

Dyck JA, Maul GG, Miller WH, Chen JD, Kakizuka A, Evans RM (1994) A novel macromolecular structure is a target of the promyelocyte-retinoic acid receptor oncoprotein. Cell 76: 333–343

Elliott S, Taylor K, White S, Rodwell R, Marlton P, Meagher D, Wiley J, Taylor D, Wright S, Timms P (1992) Proof of differentiative mode of action of all-*trans* retinoic acid in acute promyelocytic leukemia using X-linked clonal analysis. Blood 79: 1916–1919

Everett RD, Maul GG (1994) HSV-1 IE protein Vmw110 causes redistribution of PML. EMBO J 13: 5062–5069

Fagioli M, Alcalay M, Pandolfi PP, Venturini L, Mencarelli A, Simeone A, Acampora D, Grignani F, Pelicci PG (1992) Alternative splicing of PML transcripts predicts coexpression of several carboxy-terminally different protein isoforms. Oncogene 7: 1083–1091

Fenaux P, Chastang C, Castaigne S, Archimbaud E, Sanz M, Link H, Guerci A, Fegueux N, Zittoun R, Stoppa AM, Travade P, Lamy T, Maloisel F, Sadoun A, San Miguel J, Veil A, Rayon C, Conde E, Fey M, Bordessoule D, Ganser A, Bowen D, Dreyfus F, Huguet F, Tilly H, Guy H, Auzanneau G, Chomienne C, Degos L (1994) Treatment of newly diagnosed acute promyelocytic leukemia (APL) with all-transretinoic acid (ATRA) followed by intensive chemotherapy (CT). Updated results of the European group. Blood 84: (Suppl): 379a

Flenghi L, Fagioli M, Tomassoni L, Pileri S, Gambacorta M, Pacini R, Grignani F, Casini T, Ferrucci PF, Martelli MF, Pelicci P-G, Falini B (1995) Characterization of a new monoclonal antibody (PG-M3) directed against the aminoterminal portion of the PML gene product : immunocytochemical evidence for high expression of PML proteins on activated macrophages, endothelial cells, and epithelia. Blood 85: 1871–1880

Forman BM, Umesono K, Chen J, Evans RM (1995) Unique response pathways are established by allosteric interactions among nuclear hormone receptors. Cell 81: 541–550

Frankel SR, Eardley A, Lauwers G, Weiss M, Warrell RP (1992) The "retinoic acid syndrome" in acute promyelocytic leukemia. Ann Intern Med 117: 292–296

Freemont PS, Hanson IM, Trowsdale J (1991) A novel cysteine-rich sequence motif. Cell 64: 483–484

Freemont PS (1993) The RING Finger. A novel protein sequence motif related to the zinc finger. Ann NY Acad Sci 684: 174–192

Fukutani H, Naoe T, Ohno R, Yoshida H, Miyawaki S, Shimazaki C, Miyake T, Nakayama Y, Kobayashi H, Goto S, Takeshita A, Kobayashi S, Kato Y, Shiraishi K, Sasada M, Ohtake S, Murakami H, Kobayashi S, Endo N, Shindo H, Matsushita K, Hasegawa S, Tsuji K, Ueda Y, Tominaga N, Furuya H, Inoue Y, Takeuchi J, Morishita H, Iida H (1995) Isoforms of PML-retinoic acid receptor alpha fused transcripts affect neither clinical features of acute promyelocytic leukemia nor prognosis after treatment with all-trans retinoic acid. Leukemia 9: 1478–1482

Gallagher RE, Li Y-P, Rao S, Paietta E, Andersen J, Etkind P, Bennett JM, Tallman MS, Wiernik PH (1995) Characterization of acute promyelocytic leukemia cases with PML-RARα break/fusion sites in PML exon 6: identification of a subgroup with decreased in vitro responsiveness to all-*trans* retinoic acid. Blood 86: 1540–1547

Giguere V, Ong ES, Segui P, Evans RM (1987) Identification of a receptor for the morphogen retinoic acid. Nature 330: 624–629

Goddard AD, Borrow J, Freemont PS, Solomon E (1991) Characterization of a zinc finger gene disrupted by the t(15;17) in acute promyelocytic leukemia. Science 254: 1371–1374

Golomb HM, Rowley JD, Vardiman JW, Testa JR, Butler A (1980) "Microgranular" acute promyelocytic leukemia: a distinct clinical, ultrastructural and cytogenetic entity. Blood 55: 253–259

Grignani F, Ferrucci PF, Testa U, Talamo G, Fagioli M, Alcalay M, Mencarelli A, Peschle C, Nicoletti I, Pelicci PG (1993) The acute promyelocytic leukemia-specific PML-RARα fusion protein inhibits differentiation and promotes survival of myeloid precursor cells. Cell 74: 423–431

Grimwade D, Howe K, Langabeer S, Burnett A, Goldstone A, Solomon E (1996a) Minimal residual disease detection in acute promyelocytic leukemia by reverse-transcriptase PCR: evaluation of *PML-RARα* and RARα-PML assessment in patients who ultimately relapse. Leukemia 10: 61–66

Grimwade D, Howe K, Langabeer S, Davies L, Oliver F, Walker H, Swirsky D, Wheatley K, Goldstone A, Burnett A, Solomon E (1996b) Establishing the presence of the t(15;17) in suspected acute pronyelocytic leukaemia (APL); cytogenetic, molecular and PML immunofluorescence assessment of patients entered into the MRC ATRA Trial. Br J Haematol (in press)

Hiorns LR, Min T, Swansbury GJ, Zelent A, Dyer MJS, Catovsky D (1994) Interstitial insertion of retinoic acid receptor-α gene in acute promyelocytic leukaemia with normal chromosomes 15 and 17. Blood 83: 2946-2951

Hörlein AJ, Näär AM, Heinzel T, Torchia J, Gloss B, Kurokawa R, Ryan A, Kamei Y, Söderström M, Glass CK, Rosenfeld MG (1995) Ligand-independent repression by the thyroid hormone receptor mediated by a nuclear receptor co-repressor. Nature 377: 397–404

Huang M-E, Ye Y-C, Chen S-R, Chai J-R, Lu J-X, Zhoa L, Gu L-J, Wang Z-Y (1988) Use of all-transretinoic acid in the treatment of acute promyelocytic leukemia. Blood 72: 567–572

Huang W, Sun G-L, Li X-S, Cao Q, Lu Y, Jang G-S, Zhang F-Q, Chai J-R, Wang Z-Y, Waxman S, Chen Z, Chen S-J (1993) Acute promyelocytic leukemia: clinical relevance of two major PML/RARa isoforms and detection of minimal residual disease by retro-transcriptase polymerase chain reaction to detect relapse. Blood 82: 1264–1269

Imakado S, Bickenbach JR, Bundman DS, Rothnagel JA, Attar PS, Wang X-J, Walczak VR, Wisniewski S, Pote J, Gordon JS, Heyman RA, Evans RM, Roop DR (1995) Targeting expression of a dominant-negative retinoic acid receptor mutant in the epidermis of transgenic mice results in loss of barrier function. Genes Dev 9: 317–329

Jones JS, Weber S, Prakash L (1988) The *Saccharomyces cerevisiae RAD18* gene encodes a protein that contains potential zinc finger domains for nucleic acid binding and a putative nucleotide binding sequence. Nucleic Acids Res 16: 7119–7131

Kakizuka A, Miller WH, Umesono K, Warrell RP, Frankel SR, Murty VVVS, Dmitrovsky E, Evans RM (1991) Chromosomal translocation t(15;17) in human acute promyelocytic leukemia fuses RARα with a novel putative transcription factor, PML. Cell 66: 663–674

Kastner P, Perez A, Lutz Y, Rochette-Egly C, Gaub M-P, Durand B, Lanotte M, Berger R, Chambon P (1992) Structure, localisation and transcriptional properties of two classes of retinoic acid receptor α fusion proteins in acute promyelocytic leukemia (APL) : structural similarities with a new family of oncoproteins. EMBO J 11: 629–642

Kastner P, Grondona JM, Mark M, Gansmuller A, LeMeur M, Decimo D, Vonesch J-L, Dolle P, Chambon P (1994) Genetic analysis of RXRα developmental function : convergence of RXR and RAR signalling pathways in heart and eye morphogenesis. Cell 78: 987–1003

Kelly C, Vandriel R, Wilkinson GWG (1995) Disruption of PML-associated nuclear bodies during human cytomegalovirus infection. J Gen Virol 76: 2887–2893

Kliewer SA, Umesono K, Mangelsdorf DJ, Evans RM (1992) Retinoid X receptor interacts with nuclear receptors in retinoic acid, thyroid hormone and vitamin D_3 signalling. Nature : 355, 446–449

Koken MHM, Puvion-Dutilleul F, Guillemin MC, Viron A, Linares-Cruz G, Stuurman N, de Jong L, Szostecki C, Calvo F, Chomienne C, Degos L, Puvion E, de Thé H (1994) The t(15;17) translocation alters a nuclear body in a retinoic acid-reversible fashion. EMBO J 13: 1073–1083

Koken MHM, Linares-Cruz G, Quignon F, Viron A, Chelbi-Alix MK, Sobczak-Thépot J, Juhlin L, Degos L, Calvo F, de Thé H (1995) The PML growth-suppressor has an altered expression in human oncogenesis. Oncogene 10: 1315–1324

Korioth F, Gieffers C, Maul GG, Frey J (1995) Molecular characterization of NDP52, a novel protein of the Nuclear Domain 10, which is redistributed upon virus infection and interferon treatment. J Cell Biol 130: 1–13

Korninger L, Knobl P, Laczika K, Mustafa S, Quehenberger P, Schwarzinger I, Lechner K, Jaeger U, Mannhalter C (1994) PML-RARα PCR positivity in the bone marrow of patients with APL precedes haematological relapse by 2-3 months. Br J Haematol 88: 427–431

Kurokawa R, DiRenzo J, Boehm M, Sugarman J, Gloss B, Rosenfeld MG, Heyman RA, Glass CK (1994) Regulation of retinoid signalling by receptor polarity and allosteric control of ligand binding. Nature 371: 528–531

Kurokawa R, Söderström M, Hörlein A, Halachmi S, Brown M, Rosenfeld MG, Glass CK (1995) Polarity-specific activities of retinoic acid receptors determined by a co-repressor. Nature 377: 451–454

Lanotte M, Martin V, Najman S, Ballerini P, Valensi S, Berger R (1991) NB4, a maturation inducible cell-line with t(15;17) marker isolated from a human promyelocytic leukemia (M3). Blood 77: 1080–1086

Le Douarin B, Zechel C, Garnier J-M, Lutz Y, Tora L, Pierrat B, Heery D, Gronemeyer H, Chambon P, Losson R (1995) The N-terminal part of TIFI, a putative mediator of the ligand-dependent activation function (AF-2) of nuclear receptors, is fused to B-raf in the oncogenic protein T18. Embo J 14: 2020–2033

Lehmann JM, Jong L, Fanjul A, Cameron JF, Lu XP, Haefner P, Dawson MI, Pfahl M (1992) Retinoids selective for retinoid X receptor response pathways. Science 258: 1944–1946

Leid M, Kastner P, Chambon P (1992a) Multiplicity generates diversity in the retinoic acid signalling pathways. TIBS 17: 427–433

Leid M, Kastner P, Lyons R, Nakshatri H, Saunders M, Zacharewski T, Chen J-Y, Staub A, Garnier J-M, Mader S, Chambon P (1992b) Purification, cloning, and RXR identity of the HeLa cell factor with which RAR or TR heterodimerizes to bind target sequences efficiently. Cell 68: 377–395

Lemons RS, Eilender D, Waldmann RA, Robentisch M, Frej AK, Ledbetter DM, Willman C, McConnel P (1990) Cloning and characterisation of the t(15;17) translocation breakpoint region in acute promyelocytic leukemia. Genes Chromosom Cancer 2:79–87

Leroy P, Krust A, Zelent A, Mendelsohn C, Garnier J-M, Kastner P, Dierich A, Chambon P (1991) Multiple isoforms of the mouse retinoic acid receptor α are generated by alternative splicing and differential induction by retinoic acid. EMBO J 10: 59–69

Licht JD, Chomienne C, Goy A, Chen A, Scott AA, Head DR, Michaux JL, Wu Y, DeBlasio A, Miller WH, Zelenetz AD, Willman CL, Chen Z, Chen S-J, Zelent A, Macintyre E, Veil A, Cortes J, Kantarjian H, Waxman S (1995) Clinical and molecular characterization of a rare syndrome of acute promyelocytic leukemia associated with translocation (11;17). Blood 85: 1083–1094

Linch DC, Fine LG, Fielding A, Machin SJ, Goldstone AH, Solomon E, Gann E (1994) Acute promyelocytic leukaemia. Lancet 344: 1615–1618

Liu J-H, Mu Z-M, Chang KS (1995) PML suppresses oncogenic transformation of NIH/3T3 cells by activated *neu*. J Exp Med 181: 1965–1973

Lo Coco F, Diverio D, Pandolfi PP, Biondi A, Rossi V, Avvisati G, Rambaldi A, Arcese W, Petti MC, Meloni G, Mandelli F, Grignani F, Masera G, Barbui T, Pelicci PG (1992) Molecular evaluation of residual disease as a predictor of relapse in acute promyelocytic leukemia. Lancet 340: 1437–1438

Longo L, Pandolfi PP, Biondi A, Rambaldi A, Mencarelli A, Lo Coco F, Diverio D, Pegoraro L, Avanzi G, Tabilio A, Zangrilli D, Alcalay M, Donti E, Grignani F, Pelicci PG (1990) Rearrangements and aberrant expression of the retinoic acid α gene in acute promyelocytic leukemias. J Exp Med 172: 1571–1575

Longo L, Trecca D, Biondi A, Lo Coco F, Grignani F, Maiolo MA, Pelicci PG, Neri A (1993) Frequency of RAS and p53 mutations in acute promyelocytic leukemias. Leuk Lymphoma 11: 405–410

Luisi BF, Xu WX, Otwinowski Z, Freedman LP, Yamamoto KR, Sigler PB (1991) Crystallographic analysis of the interaction of the glucocorticoid receptor with DNA. Nature 352: 497–505

McKenna RW, Parkin J, Bloomfield CD, Sundberg RD, Brunning RD (1982) Acute promyelocytic leukaemia : a study of 39 cases with identification of a hyperbasophilic microgranular variant. Br J Haematol 50: 201–214

McKinney CD, Golden WL, Gemma NW, Swerdlow SH,Williams ME (1994) RARA and PML gene rearrangements in acute promyelocytic leukaemia with complex translocations and atypical features. Genes Chromosom Cancer 9: 49–56

Maslak P, Miller WH, Heller G, Scheinberg DA, Dmitrovsky E, Warrell RP (1993) CD2 expression and PML-RAR-a trancripts in acute promyelocytic leukemia. Blood 81: 1666-1667

Mermod N, O'Neill EA, Kelly TJ, Tjian R (1989) The proline-rich transcriptional activator of CTF/NF-1 is distinct from the replication and DNA binding domain. Cell 58: 741–753

Miki T, Fleming TP, Crescenzi M, Molloy CJ, Blam SB, Reynolds SH, Aaronson SA (1991) Development of a highly efficient expression cDNA cloning system: application to oncogene isolation. Proc Natl Acad Sci USA 88: 5167–5171

Miki Y, Swensen J, Shattuck-Eidens D, Futreal PA, Harshman K, Tavtigian S, et al. (1994) A strong candidate for the breast and ovarian cancer susceptibility gene *BRCA1*. Science 266: 66–71

Miller WH, Kakizuka A, Frankel SR, Warrell RP, DeBlasio A, Levine K, Evans R, Dmitrovsky E (1992) Reverse transcription polymerase chain reaction for the rearranged retinoic acid receptor α clarifies diagnosis and detects minimal residual disease in acute promyelocytic leukemia. Proc Natl Acad Sci USA 89: 2694–2698

Miller WH, Levine K, DeBlasio A, Frankel SR, Dmitrovsky E, Warrell RP (1993) Detection of minimal residual disease in acute promyelocytic leukemia by a reverse transcription polymerase chain reaction assay for the PML/RAR-a fusion mRNA. Blood 82: 1689–1694

Mitchell PJ, Tjian R (1989) Transcriptional regulation in mammalian cells by sequence-specific DNA binding proteins. Science 245: 371-378

Mitelman F (1994) Catalog of chromosomal aberrations in cancer 5th edn Liss, New York, pp: 2489–2499; 2759–2769

Morris SW, Kirstein MN, Valentine MB, Dittmer KG, Shapiro DN, Saltman DL, Look AT (1994) Fusion of a kinase gene, *ALK*, to a nucleolar protein gene, *NPM*, in non-Hodgkin's lymphoma. Science 263: 1281–1284

Mu Z-M, Chin K-V, Liu J-H, Lozano G, Chang K-S (1994) PML, a growth suppressor disrupted in acute promyelocytic leukemia. Mol Cell Biol 14: 6858-6867

Nagpal S, Saunders M, Kastner P, Durand B, Nakshatri H, Chambon P (1992) Promoter context- and response element-dependent specificity of the transcriptional activation and modulating functions of retinoic acid receptors. Cell 70: 1007–1019

Nucifora G, Larson RA, Rowley JD (1993) Persistence of the 8;21 translocation in patients with acute myeloid leukemia type M2 in long-term remission. Blood 82: 712–715

Paietta E, Andersen J, Gallagher R, Bennett J, Yunis J, Cassileth P, Rowe J, Wiernik PH (1994) The immunophenotype of acute promyelocytic leukemia (APL) : and ECOG study. Leukemia 8: 1108–1112

Pandolfi PP, Grignani F, Alcalay M, Mencarelli A, Biondi A, Lo Coco F, Grignani F, Pelicci PG (1991) Structure and origin of the acute promyelocytic leukemia myl/RARα cDNA and characterization of its retinoid-binding and transactivation properties. Oncogene 6: 1285–1292

Pandolfi PP, Alcalay M, Fagioli M, Zangrilli D, Mencarelli A, Diverio D, Biondi A, Lo Coco F, Rambaldi A, Grignani F, Rochette-Egly C, Gaube M-P, Chambon P, Pelicci PG (1992) Genomic variability and alternative splicing generate multiple PML/RARα transcripts that encode aberrant PML proteins and PML/RARα isoforms in acute promyelocytic leukaemia. EMBO J 11: 1397–1407

Perez A, Kastner P, Sethi S, Lutz Y, Reibel C, Chambon P (1993) PML/RAR homodimers: distinct DNA binding properties and heteromeric interaction with RXR. EMBO J 12: 3171–3182

Preston RA, Manolson MF, Becherer K, Weidenhammer E, Kirkpatrick D, Wright R, Jones EW (1991) Isolation and characterization of *PEP3*, a gene required for vacuolar biogenesis in *Saccharomyces cerevisiae*. Mol Cell Biol 11: 5801–5812

Puvion-Dutilleul F, Chelbi-Alix MK, Koken M, Quignon F, Puvion E, de Thé H (1995) Adenovirus infection induces rearrangements in the intranuclear distribution of the nuclear body-associated PML protein. Exp Cell Res 218: 9–16

Redner RL, Rush EA, Faas S, Rudert WA, Corey SJ (1996) The t(5;17) variant of acute promyelocytic leukemia expresses a nucleophosmin-retinoic acid receptor fusion. Blood 87: 882–886

Robertson KA, Emami B, Collins SJ (1992a) Retinoic acid-resistant HL-60R cells harbor a point mutation in the retinoic acid receptor ligand-binding domain that confers dominant negative activity. Blood 80: 1885–1889

Robertson KA, Emami B, Mueller L, Collins SJ (1992b) Multiple members of the retinoic acid receptor family are capable of mediating the granulocytic differentiation of HL-60 cells. Mol Cell Biol 12: 3743–3749

Robinson JS, Graham TR, Emr SD (1991) A putative zinc finger protein, *Saccharomyces cerevisiae* Vps18p, affects late Golgi functions required for vacuolar protein sorting and efficient α-factor prohormone maturation. Mol Cell Biol 12: 5813–5824

Rodeghiero F, Castaman G (1994) The pathophysiology and treatment of hemorrhagic syndrome of acute promyelocytic leukemia. Leukemia 8 (Suppl) : s20–s26

Rowley JD, Golomb HM, Dougherty C (1977) 15/17 translocation, a consistent chromosomal change in acute promyelocytic leukaemia. Lancet 1: 549–550

Saitou M, Sugai S, Tanaka T, Shimouchi K, Fuchs E, Narumiya S, Kakizuka A (1995) Inhibition of skin development by targeted expression of a dominant negative retinoic acid receptor. Nature 374: 159–162

Sakashita A, Kizaki M, Pakkala S, Schiller G, Tsuruoka, Tomosaki R, Cameron JF, Dawson MI, Koeffler HP (1993) 9-*cis*-retinoic acid : effects on normal and leukaemic haematopoiesis in vitro. Blood 81:1009–1016

Sarkar A, Yang P, Fan Y-H, Mu Z-M, Hauptmann R, Adolf GR, Stass SA, Chang K-S (1994) Regulation of the expression of annexin VIII in acute promylocytic leukemia. Blood 84: 279–286

Sawyers CL, Denny CT, Witte ON (1991) Leukemia and the disruption of normal haematopoiesis. Cell 64: 337–350

Schatz DG, Oettinger MA, Baltimore D (1989) The V(D)J recombination activating gene, RAG-1. Cell 59: 1035–1048

Schrader M, Wyss A, Sturzenbecker LJ, Grippo JF, LeMotte P, Carlberg C (1993) RXR-dependent and RXR-independent transactivation by retinoic acid receptors. Nucleic acid Res 21: 1231–1237

Schwabe JWR, Chapman L, Finch JT, Rhodes D (1993) The crystal structure of the estrogen receptor DNA-binding domain bound to DNA : How receptors discriminate between their response elements. Cell 75: 567–578

Seale J, Marcus R, Janossy G, Swirsky D, Goldman J, Cross N, Pandolfi PP (1994) Quantitative assessment of minimal residual disease in acute promyelocytic leukemia by RT-PCR: Evaluation of techniques to improve sensitivity. Blood 84 (Suppl) : 301a

Stadler M, Chelbi-Alix MK, Koken MHM, Venturini L, Lee C, Säib A, Quignon F , Pelicano L, Guillemin M-C, Schindler C, de Thé H (1995) Transcriptional induction of the PML growth suppressor gene by interferons is mediated through an ISRE and a GAS element. Oncogene 11: 2565–2573

Stone RM, Mayer RJ (1990) The unique aspects of acute promyelocytic leukemia. J Clin Oncol 8: 1913–1921

Stunnenberg HG (1993) Mechanisms of transactivation by retinoic acid receptors. Bioessays 15: 309–315

Stuurman N, de Graaf A, Floore A, Josso A, Humbel B, deJong L, van Driel R (1992) A monoclonal antibody recognizing nuclear matrix-associated nuclear bodies. J Cell Sc 101: 773–784

Sucov HM, Dysøn E, Gumeringer CL, Price J, Chien KR, Evans RM, (1994) RXRα mutant mice establish a genetic basis for vitamin A signalling in heart morphogenesis. Genes Dev 8: 1007–1018

Swansbury GJ, Lawler SD, Alimena G, Arthur D, Berger R, Van Den Berghe H, Bloomfield CD, de la Chapelle A, Dewald G, Garson OM, Hagemeijer A, Mitelman F, Rowley JD, Sakurai M (1994) Long-term survival in acute myelogenous leukemia: A second follow-up of the fourth international workshop on chromosomes in leukemia. Cancer Genet Cytogenet 73: 1–7

Takahashi M, Inaguma Y, Hiai H, Hirose F (1988) Developmentally regulated expression of a human finger-containing gene encoded by the 5′ half of the ret transforming gene. Mol Cell Biol 8: 1853–1856

Tallman MS, Kwaan HC (1992) Reassessing the hemostatic disorder associated with acute promyelocytic leukemia. Blood 79: 543–553

Tashiro S, Kotomura N, Tanaka K, Suzuki K, Kyo T, Dohy H, Niwa O, Kamada N (1994) Identification of illegitimate recombination hot spot of the retinoic acid receptor alpha gene involved in 15;17 chromosomal translocation of acute promyelocytic leukemia. Oncogene 9: 1939–1945

Terris B, Baldin V, Dubois S, Degott C, Flejou J-F, Henin D, Dejean A (1995) PML nuclear bodies are general targets for inflammation and cell proliferation. Cancer Res 55: 1590–1597

Testa U, Grignani F, Barberi T, Fagioli M, Masciulli R, Ferruci PF, Seripa D, Camagna A, Alcalay M, Pelicci PG, Peschle C (1994) PML/RARα$^+$ U937 mutant and NB4 cell lines : retinoic acid restores the monocytic differentiation response to vitamin D_3. Cancer Res 54: 4508–4515

Tobal K , Liu Yin JA (1995) Detection of persistent PML-RARα positive cells in long term remission of acute promyelocytic leukaemia (APL) by a sensitive "hot start" RT-PCR. Br J Haematol 89(Suppl) : 59

Tsai S, Collins SJ (1993) A dominant retinoic acid receptor blocks neutrophil differentiation at the promyelocyte stage. Proc Natl Acad USA 90: 7153–7157

Umesono K, Murakami KK, Thompson CC, Evans RM (1991) Direct repeats as selective response elements for the thyroid hormone, retinoic acid, and vitamin D_3 receptors. Cell 65: 1255–1266

Vahdat L, Maslak P, Miller WH, Eardley A, Heller G, Scheinberg DA, Warrell RP (1994) Early mortality and the retinoic acid syndrome in acute promyelocytic leukemia: Impact of leukocytosis, low-dose chemotherapy, PML-RAR-α isoform, and CD13 expression in patients treated with all-*trans* retinoic acid. Blood 84: 3843–3849

Warrell RP (1993) Retinoid resistance in acute promyelocytic leukemia: new mechanisms, strategies, and implications. Blood 82: 1949–1953

Weis K, Rambaud S, Lavau C, Jansen J, Carvalho T, Carmo-Fonseca M, Lamond A, Dejean A (1994) Retinoic acid regulates aberrant nuclear localization of PML-RARα in acute promyelocytic leukaemia cells. Cell 76: 345–356

Woolford CA, Dixon CK, Manolson MF, Wright R, Jones EW (1990) Isolation and characterization of *PEP5*, a gene essential for vacuolar biogenesis in *saccharomyces cerevisiae.* Genetics 125: 739–752

Zhang X-K, Hoffmann B, Tran P B-V, Graupner G, Pfahl M (1992) Retinoid X receptor is an auxiliary protein for thyroid hormone and retinoic acid receptors. Nature 355: 441–446

Fusions Involving Paired Box and Fork Head Family Transcription Factors in the Pediatric Cancer Alveolar Rhabdomyosarcoma

F.G. BARR

1 Introduction

Rhabdomyosarcoma (RMS) is a soft tissue tumor with striated muscle differentiation which usually presents in pediatric patients (RANEY et al. 1993). Though histopathologic classification has evolved over time, two major forms of childhood RMS are consistently recognized, embryonal (ERMS) and alveolar (ARMS). These two histopathologic entities have distinct clinical behaviors. ERMS presents mainly in children less than 10 years old; occurs predominantly in the head and neck, genitourinary tract, and retroperitoneum; and is associated with a favorable prognosis. In contrast, ARMS presents mainly in adolescents and young adults, often occurs in the extremities, and is associated with an unfavorable prognosis (TSOKOS et al. 1992). This unfavorable prognosis is related to the propensity for early and wide dissemination, often involving bone marrow, and to the poor response to chemotherapy.

Department of Pathology and Laboratory Medicine, University of Pennsylvania School of Medicine, Philadelphia, PA 19104-6082, USA

Diagnosis of RMS is often complicated by a paucity of features of striated muscle differentiation (Tsokos 1994). Furthermore, the histologic criteria for distinguishing the ERMS and ARMS subtypes are relatively subtle. The problem is compounded by the fact that a variety of pediatric solid tumors including RMS, neuroblastoma, Ewing's sarcoma, and non-Hodgkin's lymphoma can present as collections of poorly differentiated cells. To detect more subtle evidence of muscle differentiation, these tumors are stained with immunohistochemical reagents specific for muscle proteins (such as MyoD, desmin, myoglobin, or muscle-specific actin) or examined by electron microscopy for myofilaments. However, there is no well-established ultrastructural or immunohistochemical marker that will distinguish ERMS and ARMS tumors.

Cytogenetic analysis of numerous RMS cases has demonstrated nonrandom chromosomal translocations associated with the ARMS subtype. The most prevalent finding is a translocation involving chromosomes 2 and 13, t(2;13)(q35-37;q14) (Fig. 1), which was detected in approximately 70% of published ARMS

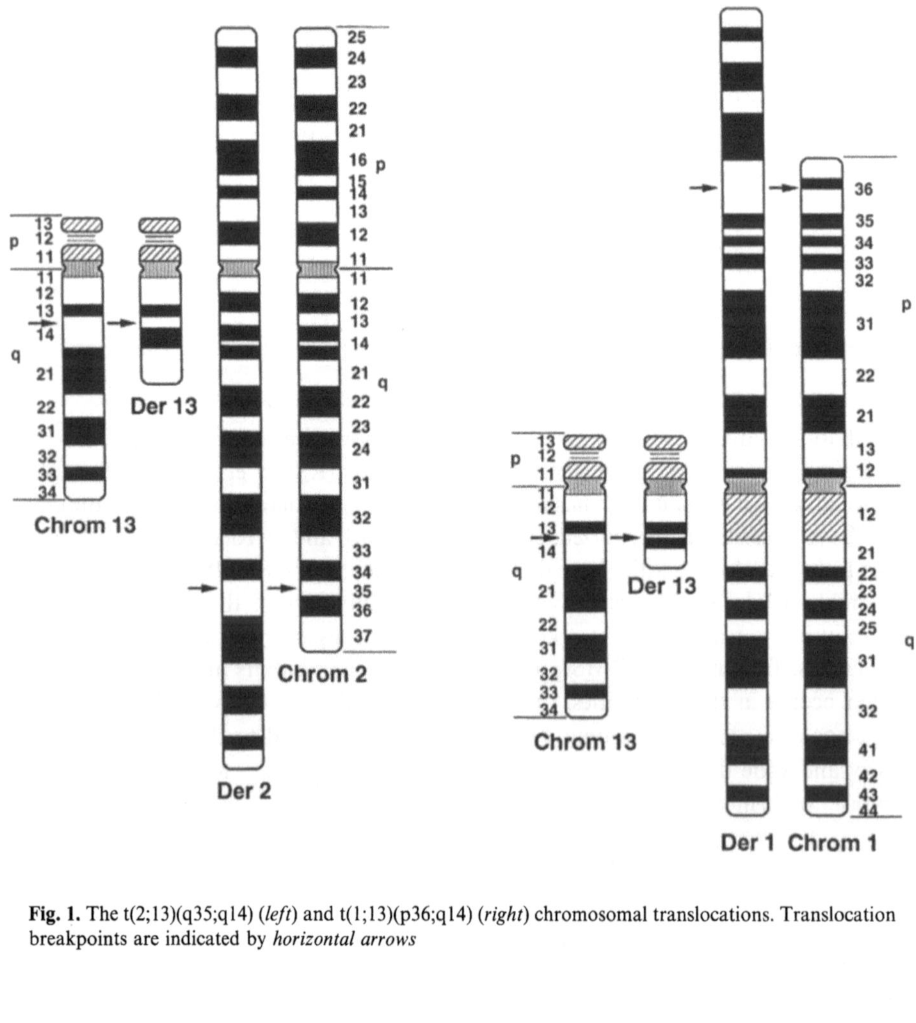

Fig. 1. The t(2;13)(q35;q14) (*left*) and t(1;13)(p36;q14) (*right*) chromosomal translocations. Translocation breakpoints are indicated by *horizontal arrows*

cases (TURC-CAREL et al. 1986; DOUGLASS et al. 1987; WANG-WUU et al. 1988; WHANG-PENG et al. 1992). In addition, there have been several reports of a t(1;13)(p36;q14) variant translocation (Fig. 1) (BIEGEL et al. 1991; DOUGLASS et al. 1991; WHANG-PENG et al. 1992), as well as single reports of cases with other related alterations (WHANG-PENG et al. 1986; DOUGLASS et al. 1991; SAWYER et al. 1994). The t(2;13) and t(1;13) translocations have not been associated with any other tumor and thus appear to be specific markers for ARMS. Though no consistent chromosomal alterations have been associated with ERMS (HEIM and MITELMAN 1995), loss of heterozygosity (SCRABLE et al. 1989) and chromosome transfer studies (KOI et al. 1993) have indicated the presence of a putative tumor suppressor gene in chromosomal region 11p15, inactivation of which is involved in development of ERMS but not ARMS.

In summary, clinical and basic research investigations of RMS have demonstrated significant problems from diagnostic, therapeutic, and biological perspectives. This review will focus on the molecular genetics of ARMS and specifically address the importance of the characteristic chromosomal translocations in investigating these diverse perspectives. Elucidation of the molecular basis of these chromosomal translocations has provided a valuable opportunity to investigate an important step in ARMS pathogenesis and to explore the utility of these genetic alterations as diagnostic markers or therapeutic targets in clinical management.

2 Genomic Mapping and Cloning of Genes Disrupted by the t(2;13) Translocation

Several laboratories developed strategies to localize the t(2;13) breakpoint on physical maps of chromosomes 2 and 13 (Fig. 2). For chromosome 13 studies, a panel of somatic cell hybrid and lymphoblast cell lines with constitutional deletions and unbalanced translocations involving chromosome 13 was assembled (BARR et al. 1991). Probes from the 13q12–14 region were situated by Southern blot analysis within physical intervals delimited by the constitutional breakpoints in genomic DNA isolated from these cell lines. In addition, the probes were localized with respect to the t(2;13) breakpoint in ARMS cell lines or somatic cell hybrids containing one of the translocated chromosomes (VALENTINE et al. 1989; BARR et al. 1991; MITCHELL et al. 1991; SHAPIRO et al. 1992). These experiments localized the t(2;13) breakpoint within a physical interval defined by the proximal deletion breakpoints in lymphoblast lines GM01484 and GM07312 (Fig. 2). From a comparison of the physical data with the genetic linkage map of human chromosome 13 (BOWCOCK et al. 1991), the t(2;13) breakpoint is most closely flanked by loci *D13S29* and *TUBBP2*. Using the available linkage and other mapping data, the distance between these loci is estimated to be less than 9 cM.

The t(2;13) breakpoint was sublocalized on a physical map of chromosome 2 by similar strategies (BARR et al. 1992). These experiments localized the t(2;13)

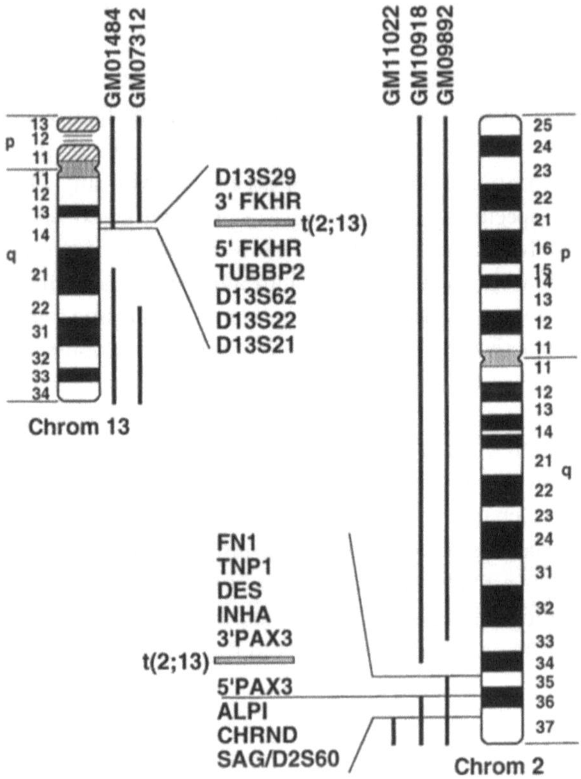

Fig. 2. Localization of t(2;13) translocation breakpoints by physical mapping analyses of chromosome 13 (*left*) and chromosome 2 (*right*). Names of cell lines with constitutional deletions or unbalanced translocations of chromosome 2 or 13 are shown at the *top* of each figure. A schematic representation of the rearranged chromosome in each cell line is shown as *vertical bars* adjacent to the chromosome diagrams. The breakpoints in these cell lines are extended by *horizontal lines* to identify small physical mapping intervals. Chromosome 2 and 13 loci were localized to the appropriate interval by Southern blot analysis of the cell lines (Barr et al. 1991; Barr et al. 1992). The *shaded bar* indicates the position of the t(2;13) translocation breakpoint determined by Southern blot and PCR assays of alveolar rhabdomyosarcoma cell lines and somatic cell hybrids derived from these tumor lines. The order of the loci within these physical intervals has been determined by linkage analysis (Bowcock et al. 1991; Malo et al. 1991)

breakpoint to a physical interval delimited by the distal deletion breakpoint in lymphoblast line GM09892 and the t(X;2) breakpoint in somatic cell hybrid GM11022 (Fig. 2). These results demonstrated a physical order of human loci consistent with the order of loci derived from genetic linkage analysis of human chromosome 2 (O'Connell et al. 1989) and the linkage order in the syntenic region on mouse chromosome 1 (Malo et al. 1991; Watson et al. 1992). The linkage data predicted that the human loci that most closely flank the t(2;13) breakpoint are *INHA* and *ALPI*. The distance between the murine counterparts on mouse chromosome 1 has been calculated to be 5.7–13.4 cM.

Linkage studies of mouse chromosome 1 have indicated that *Pax-3*, which encodes a member of the paired box transcription factor family (Goulding et al.

1991), maps between the mouse homologues of *INHA* and *ALPI* (Epstein et al. 1991). Using a chromosome 2 mapping panel, the human homologue (termed *PAX3* or *HuP2*) (Burri et al. 1989) was localized to the human 2q region adjacent to the t(2;13) breakpoint (Fig. 2) (Barr et al. 1993). Somatic cell hybrids derived from an ARMS were then analysed by polymerase chain reaction (PCR) with oligonucleotide primers specific for the 5′ and 3′ ends of *PAX3*. These PCR studies indicated that *PAX3* is split by the t(2;13) such that the 5′ PAX3 region is located on the derivative chromosome 13 [der(13)] and the 3′ PAX3 region is on the derivative chromosome 2 [der(2)]. Structural alterations of *PAX3* were confirmed by Southern blot analysis of genomic DNA from ARMS cell lines. Therefore, *PAX3* is the chromosome 2 locus rearranged by the t(2;13). The rearrangements in three ARMS cell lines were then localized to the 3′ end of the *PAX3* gene. Fine mapping of this 3′ region demonstrated that the final three exons are contained within a 40 kb region. The translocation breakpoints in the t(2;13)-containing cell lines are situated within a 20 kb intron between the final and penultimate exons.

Northern analysis with a PAX3 probe demonstrated a novel 7.2 kb transcript unique to the cell lines with the t(2;13) (Barr et al. 1993; Shapiro et al. 1993a). This transcript only hybridizes to PAX3 sequences 5′ to the t(2;13) breakpoint indicating that it is the transcription product of the rearranged *PAX3* gene located on the der(13). To further characterize this transcript, clones containing the expected PAX3 cDNA sequence 5′ to the t(2;13) breakpoint fused to a novel sequence (Fig. 3) were isolated from cDNA libraries constructed from ARMS cell lines (Barr et al. 1993; Shapiro et al. 1993a). This novel sequence was localized on chromosome 13 to the region that contains the t(2;13) breakpoint (Fig. 2). Northern blot analysis of

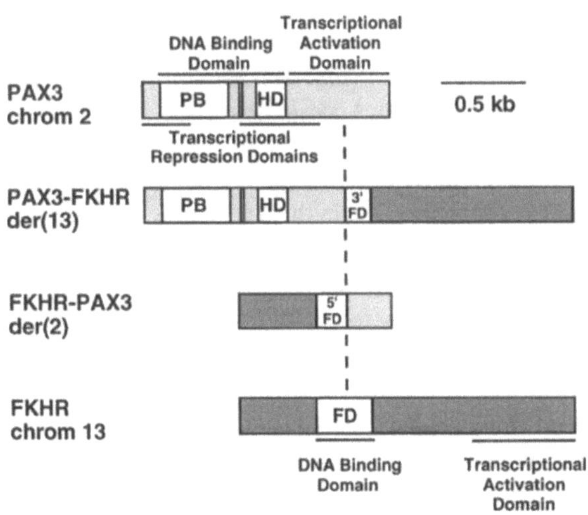

Fig. 3. Maps of the wild-type and chimeric coding regions associated with the t(2;13) translocation. In PAX3, the conserved paired box (PB), octapeptide, and homeodomain (HD), and in FKHR, the conserved fork head domain (FD) are shown as *open boxes*. The fusion point is marked by a *vertical dashed line*. Functional domains are shown as *solid horizontal bars*

ARMS cells with this sequence demonstrated hybridization to the same novel band detected by the 5′ PAX3 probe. These findings indicate that the t(2;13) results in a chimeric transcript composed of 5′ PAX3 sequences fused to sequences from a chromosome 13q14 gene.

A full-length cDNA from the wild-type chromosome 13 gene was isolated as overlapping clones from several libraries (GALILI et al. 1993; SHAPIRO et al. 1993a). This cDNA detected a 6.5 kb transcript in lymphoblasts, fibroblasts, and numerous fetal and adult tissues, and thus corresponds to a widely expressed gene. Sequence analysis revealed a 1965 bp open reading frame (ORF) encoding a 655 amino acid protein (Fig. 3). A BLAST database search showed homology to the fork head or winged-helix transcription factor family, and thus the chromosome 13 gene has been named *FKHR* (fork head in rhabdomyosarcoma) (GALILI et al. 1993). This gene has also been referred to as *ALV* (SHAPIRO et al. 1993a).

3 PAX3 and the Paired Box Transcription Factor Family

The *PAX3* gene is a member of the paired box or PAX transcription factor family which is characterized by the highly conserved paired box DNA binding domain that was first identified in *Drosophila* segmentation genes (STRACHAN and READ 1994; TREMBLAY and GRUSS 1994). Nine human members of this family have been subsequently identified. Several of these genes also contain a complete or truncated version of a second conserved DNA binding domain, the homeobox, as well as a short conserved octapeptide motif distal to the paired box. The *PAX3* gene encodes a protein containing an NH_2-terminal DNA binding domain consisting of a paired box, octapeptide, and complete homeodomain, and a COOH-terminal serine- and threonine-rich transcriptional activation domain (Fig. 3) (GOULDING et al. 1991). An identical organization and highly homologous coding sequence are also found in the *PAX7* gene (JOSTES et al. 1991; DAVIS et al. 1994; SCHÄFER et al. 1994), and thus *PAX3* and *PAX7* constitute a subfamily within the paired box family (Fig. 4).

The paired box family is postulated to function in the transcriptional control of development (STRACHAN and READ 1994; TREMBLAY and GRUSS 1994). Each of the genes has a specific temporal and spatial pattern of expression during early development, and some are also expressed with a very restricted distribution in the adult. In situ hybridization analysis of murine embryos has revealed PAX3 expression in the developing nervous system as well as the dermomyotome and limb bud mesenchyme, which contain skeletal muscle precursors (GOULDING et al. 1991). In contrast, PAX3 expression was not detected in the adult mouse. Other studies have demonstrated that PAX7 is expressed during murine embryogenesis in similar areas but is activated later and persists longer than that of PAX3 (JOSTES et al. 1991).

The important developmental role of the paired box genes is further emphasized by the association of mutations in several of the paired box genes with heritable murine and human developmental defects (STRACHAN and READ 1994;

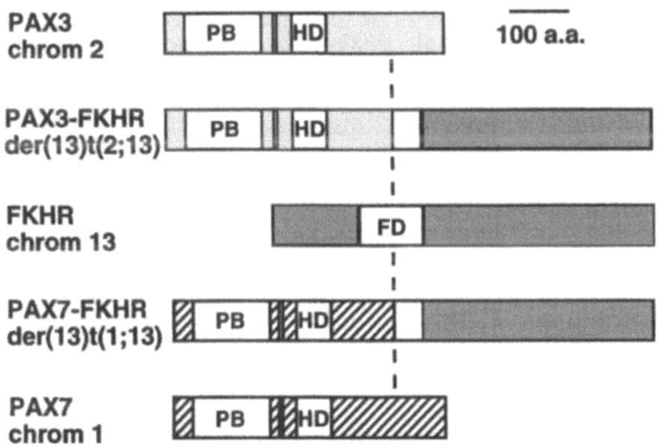

Fig. 4. Comparison of wild-type and fusion products associated with the t(2;13) and t(1;13) translocations. The conserved motifs are shown as *open boxes* as described in Fig. 3. The fusion points are marked by a *vertical dashed line*

TREMBLAY and GRUSS 1994). Point mutations and deletions affecting the DNA binding domain of the *Pax-3* gene have been identified in the *splotch* mouse (EPSTEIN et al. 1991; STRACHAN and READ 1994; TREMBLAY and GRUSS 1994), which is characterized by abnormalities of the neural tube, neural crest-derived structures, and peripheral musculature. The human disease Waardenburg syndrome, characterized by deafness and pigmentary disturbances, is also caused by mutations affecting the functional domains of the *PAX3* gene (BALDWIN et al. 1992; TASSABEHJI et al. 1992; STRACHAN and READ 1994; TREMBLAY and GRUSS 1994). Mutations causing *splotch* and Waardenburg syndrome appear to alter or abolish the DNA binding activity (CHALEPAKIS et al. 1994a; UNDERHILL et al. 1995) and therefore result in a loss of function during important periods of development.

4 FKHR and the Fork Head Transcription Factor Family

The fork head domain was initially identified as an approximately 100 amino acid region of sequence similarity between the *Drosophila* homeotic gene *fork head* and the rat gene encoding the hepatocyte transcription factor HNF-3α (WEIGEL et al. 1989; LAI et al. 1990; WEIGEL and JACKLE 1990). During the past few years, over 30 genes which share this conserved motif have been isolated from species ranging from yeast to human (LAI et al. 1993). Several studies have demonstrated that the fork head domain is necessary and sufficient for DNA binding activity, whereas other divergent regions of the proteins confer transcriptional regulatory function. Crystallographic analysis of the HNF-3γ fork head domain revealed an arrangement of two long loops attached to a compact core of three α-helices; this arrangement has been named the winged-helix (CLARK et al. 1993).

Comparison of *FKHR* with other members of the fork head family suggest that it is a relatively divergent member (Galili et al. 1993). The fork head domain of FKHR (Fig. 3) lacks the NH_2-terminal KPPY common to most fork head genes (Clevidence et al. 1993) and contains a novel five amino acid insert in the middle of the motif. FKHR does not show homology with family members outside the fork head domain. In the COOH-terminal, there is an acidic region which functions as a transcriptional activation domain (Fig. 3) (Bennicelli et al. 1995). Other motifs in FKHR include a ten amino acid NH_2-terminal proline-rich sequence similar to a potential SH3 binding site (Ren et al. 1993) and an NH_2-terminal alanine-rich sequence which has been associated in other proteins with transcriptional repression (Han and Manley 1993).

Genetic, expression, and functional analyses of fork head genes have indicated diverse roles in control of embryonic development and adult tissue-specific gene expression (Lai et al. 1993). In the immunodeficient *nude* phenotype of mice and rats, mutations disrupting the DNA binding domain have been identified in a fork head gene termed *whn* (Nehls et al. 1994). Identification of the avian retroviral oncogene *qin* as a fork head gene has also raised the possibility of growth-regulatory function (Li and Vogt 1993). In addition, the fork head family member *AFX1* has an oncogenic role as indicated by the finding of fusion of *AFX1* to the *MLL* gene by a t(X;11)(q13;q23) chromosomal translocation in acute leukemia (Parry et al. 1994).

5 PAX3-FKHR Chimeric Transcripts and Proteins in Alveolar Rhabdomyosarcoma

Sequence analysis of the chimeric PAX3-FKHR cDNA cloned from an ARMS cell line revealed that the 5′ PAX3 and 3′ FKHR coding sequences are fused in-frame (Galili et al. 1993; Shapiro et al. 1993a). The cDNA contains an open reading frame of 2508 nt encoding an 836 amino acid fusion protein (Fig. 3). The FKHR breakpoint occurs within the fork head domain, whereas the PAX3 breakpoint occurs distal to the homeodomain. Therefore, this fusion protein contains an intact PAX3 DNA binding domain, the COOH-terminal half of the fork head domain, and the COOH-terminal FKHR region. The consistency of this chimeric transcript was confirmed by reverse transcriptase (RT)-PCR experiments with oligonucleotide primers specific for the 5′ PAX3 and 3′ FKHR sequences (Galili et al. 1993; Shapiro et al. 1993a). The predicted PCR product was detected in eight of eight ARMS cell lines, including one line without a cytogenetically identifiable t(2;13) translocation. Sequence analysis of several PCR products confirmed the presence of the PAX3-FKHR fusion and demonstrated an invariant fusion point (Shapiro et al. 1993a; Davis et al. 1994; Downing et al. 1995). Together with the previous finding that the chromosome 2 breakpoints consistently occur in the same PAX3 intron, these data suggest that the chromosome 13 breakpoints occur within a single FKHR intron, resulting in a consistent fusion of 5′ PAX3 to 3′ FKHR exons.

The reciprocal translocation product, the der(2) chimeric transcript consisting of 5′ FKHR and 3′ PAX3 exons, was not detected in ARMS lines by northern blot analysis (BARR et al. 1993). Using a more sensitive RT-PCR assay, a 5′ FKHR-3′ PAX3 fusion was detected in six of eight ARMS lines (GALILI et al. 1993). Southern blot analysis of these PCR products confirmed that a 5′ FKHR-3′ PAX3 fusion was expressed in a subset of the ARMS lines. Thus, a spliced chimeric transcript can be expressed from the der(2) and encodes a protein consisting of the NH_2-terminal of FKHR, the NH_2-terminal half of the fork head domain, and the COOH-terminal PAX3 region (Fig. 3). Although the 5′ FKHR-3′ PAX3 fusion is expressed in some tumors, the findings of higher and more consistent expression of the 5′ PAX3-3′ FKHR fusion (BARR et al. 1993; GALILI et al. 1993) suggest that the der(13) encodes the product involved in the pathogenesis of ARMS.

To identify the fusion protein, polyclonal antisera specific for the PAX3 and FKHR proteins were prepared (GALILI et al. 1993; FREDERICKS et al. 1995). Immunoprecipitation of proteins from t(2;13)-containing ARMS cell extracts showed that both antisera detected a 97 kDa protein which was not present in t(2;13)-negative cells. The molecular mass of 97 kDa agrees well with the predicted 837 amino acid PAX3-FKHR fusion protein. Sequential immunoprecipitation confirmed that both PAX3- and FKHR-specific antisera detect the same 97 kDa polypeptide, showing conclusively that a PAX3-FKHR fusion protein is produced in ARMS. In accord with RNA expression results, an FKHR-PAX3 protein corresponding to the der(2) product was not detectably expressed.

6 PAX7-FKHR Fusion Resulting from Variant t(1;13) Translocation

Though the t(2;13)(q35;q14) translocation was found in most cases of ARMS, several cases were reported with a variant t(1;13)(p36;q14) translocation (Fig. 1) (BIEGEL et al. 1991; DOUGLASS et al. 1991; WHANG-PENG et al. 1992). In these t(1;13) cases, a PAX3-FKHR fusion was not detected by RT-PCR analysis (DAVIS et al. 1994), indicating that the t(1;13) is not a complex translocation masking involvement of chromosomal region 2q. An alternative possibility is that the t(1;13) results from juxtaposition of *FKHR* with a gene from chromosome 1. *PAX7*, another member of the paired box-containing transcription factor family, was localized to chromosomal region 1p36 (SCHÄFER et al. 1993; SHAPIRO et al. 1993b; STAPLETON et al. 1993). Southern blot and RT-PCR analyses indicated that *PAX7* is rearranged in the t(1;13)-containing tumors and fused to *FKHR* on chromosome 13 (DAVIS et al. 1994). This fusion results in a chimeric transcript consisting of 5′ PAX7 and 3′ FKHR regions, which is similar to the 5′ PAX3-3′ FKHR transcript formed by the t(2;13) translocation (Fig. 4). The 5′ PAX3 and PAX7 regions encode highly related DNA binding domains (Goulding et al. 1991; Jostes et al. 1991; Davis et al. 1994), and therefore these translocations are postulated to create similar chimeric transcription factors that alter expression of a common group of target genes.

Multiple tumors have been subsequently studied by RT-PCR analyses to determine the frequency of PAX3-FKHR and PAX7-FKHR fusions in ARMS and ERMS. One study has detected PAX3-FKHR and PAX7-FKHR fusions in 76% and 10%, respectively, of 21 histologically diagnosed ARMS cases (Barr et al. 1995). In addition, this study detected fusions in 3% and 3%, respectively, of 30 histologically diagnosed ERMS cases. Another study found PAX3-FKHR fusions in 87% of 23 ARMS tumors and 17% of 12 ERMS tumors (Downing et al. 1995). These findings indicate that there is substantial but not perfect overlap between molecular and histologic analyses and that there is a small subset of ARMS tumors which express neither PAX3-FKHR nor PAX7-FKHR fusions. The possibility of rare variant fusions or other genetic events that can substitute for the characteristic fusions should therefore be considered.

7 DNA Binding Properties of Wild-Type PAX3 and PAX3-FKHR Fusion Proteins

The previously described analyses of the PAX3-FKHR fusion indicate that the PAX3 paired box and homeodomain remain intact whereas the FKHR fork head domain is split and only the COOH-terminal half is present in the fusion product (Fig. 3) (Galili et al. 1993; Shapiro et al. 1993). Since mutations of other fork head domains have been shown to inactivate DNA binding function (Lai et al. 1990; Clevidence et al. 1993), the truncated fork head domain in the PAX3-FKHR fusion is probably inactive or serves to modify DNA binding activity of the PAX3 domains. Therefore, the PAX3 DNA binding domain is postulated to provide the DNA binding specificity for the fusion transcription factor.

Though a few physiological targets of other paired box proteins have been identified in mammalian cells (Strachan and Read 1994; Tremblay and Gruss 1994), no mammalian gene target for PAX3 binding has yet been identified. However, the PAX3 protein has been shown to bind a sequence called e5 (Goulding et al. 1991), which was identified in the *Drosophila even-skipped* gene promoter as a binding site for the *Drosophila paired* gene product, which also contains a paired box and homeodomain (Treisman et al. 1991). Footprinting experiments with e5 and its derivatives have shown that the paired and PAX3 proteins protect a 18-27 bp region, which can be subdivided into paired box- and homeodomain-specific binding sites (Treisman et al. 1991; Chalepakis et al. 1994a). Optimal binding of PAX3 to the e5 sequence requires both of these sites. Binding experiments with mutant proteins and e5 derivatives have indicated functional interaction between the paired box and homeodomain within the wild-type PAX3 protein (Chalepakis et al. 1994a; Underhill et al. 1995).

Other experiments have recently demonstrated PAX3 binding to other sequences. The PAX3 protein, as well as several other paired box proteins, bind in vitro to a subset of the sequences recognized by the PAX5 protein (CZERNY et al. 1993). These sequences are related to the 3′ portion of a bipartite consensus binding site for PAX5 and are postulated to be recognized by the NH_2-terminal portion of the paired box. Comparison of PAX3 and PAX7 with a variety of these binding sites demonstrate very comparable binding activity (SCHÄFER et al. 1994), thereby supporting the hypothesis that these transcription factors recognize a similar set of target genes.

A set of binding sites for the paired domain of PAX3 was isolated from a pool of random oligonucleotides using a PCR-based selection strategy (EPSTEIN et al. 1994b). The consensus binding site sequence is similar to that determined for the paired domains of PAX2 and PAX6 (EPSTEIN et al. 1994a). A sequence highly homologous to the PAX3 paired domain binding consensus was subsequently located in the 3′ untranslated regions of the *NFI* gene and demonstrated tight binding to PAX3 protein by in vitro binding assay (EPSTEIN et al. 1994b; EPSTEIN et al. 1995).

The DNA binding properties of the PAX3-FKHR fusion protein were assayed by electrophoretic mobility shift experiments with an oligonucleotide containing the e5 site (FREDERICKS et al. 1995). Experiments using in vitro translated protein, extracts of transfected COS-1 cells, or extracts of ARMS cells demonstrated binding to e5. The specificity of the protein-DNA complexes was shown by disappearance of shifted bands when excess unlabeled e5 or PAX3-specific antiserum was added to the incubation step, but not when unrelated oligonucleotides or preimmune antiserum was added. Comparison of the autoradiographic intensities of the protein-DNA complexes formed with PAX3 and PAX3-FKHR suggested that the two proteins may not bind to e5 with equal affinity. To further evaluate this issue, experiments were conducted with a range of concentrations of the wild-type and fusion proteins. Measurement of the intensities of the band shifts demonstrated a linear relationship between input protein and bound fragment. From the slopes of these plots, the binding affinity of the wild-type PAX3 protein for the e5 site was calculated to be 3.5-fold greater than that of the PAX3-FKHR fusion. Therefore, even though the wild-type and fusion proteins contain the same PAX3 DNA binding domain, the binding function is dependent on the protein context.

8 Transcriptional Regulatory Function of Wild-Type PAX3 and PAX3-FKHR Fusion Proteins

In the conversion from the wild-type PAX3 protein to the PAX3-FKHR fusion protein, the COOH-terminal PAX3 region is replaced with the COOH-terminal FKHR region (Fig. 3) (GALILI et al. 1993; SHAPIRO et al. 1993). To investigate the transcriptional regulatory activities of these regions independent of their respective

DNA binding domains, the COOH-terminal regions were examined as fusions to the GAL4 DNA binding domain (Chalepakis et al. 1994b; Bennicelli et al. 1995). Each GAL4 fusion construct activated transcription of a GAL4-dependent reporter gene in NIH 3T3 cells with an activity one to two orders of magnitude greater than baseline activity present in negative controls and the same order of magnitude as that of a GAL4-VP16 fusion (Sadowski et al. 1988). Deletion mapping demonstrated essential *trans*-activation domains in the extreme 3′ portions of the PAX3 and FKHR coding regions. The essential PAX3 *trans*-activation domain is serine- and threonine-rich, whereas the essential FKHR *trans*-activation domain contains both acidic and serine-, threonine-rich regions. In addition to these essential domains, positive modifying elements were identified in adjacent 5′ coding sequences. These data demonstrate that PAX3 and PAX3-FKHR contain potent, yet structurally distinct transcriptional activation domains which are switched by the t(2;13) translocation in ARMS.

The preceding experiments indicate that the PAX3-FKHR fusion protein contains a functional DNA binding domain and a potent transcriptional activation domain. To analyze the function of the wild-type PAX3 and PAX3-FKHR fusion proteins as sequence-specific transcriptional activators, constructs expressing the full-length proteins were transfected into mammalian cells along with a reporter construct containing PAX3 binding sites. In one set of experiments in NIH 3T3 cells (Fredericks et al. 1995), expression of the wild-type PAX3 protein resulted in a low but significant level of transcriptional activation from one or four tandemly repeated e5 sites. The PAX3-FKHR fusion protein induced much higher levels of transcriptional activity: 3.5-fold more activity from a single e5 site and greater than ten fold more activity from a multimer of four e5 sites. In a second set of experiments using P19 embryonal carcinoma cells and a binding site from the 3′ region of the *NFI* gene (Epstein et al. 1995), transcriptional activation by PAX3-FKHR was only 30% higher than that by PAX3. These experiments demonstrate that the PAX3-FKHR fusion protein can function as a transcription factor, specifically activating expression of genes containing PAX3 DNA binding sites. Furthermore, the fusion protein is a more potent transcriptional activator than the wild-type PAX3 protein, though the increased potency may be cell type- or binding site-specific.

GAL4-fusion experiments involving the NH_2-terminal portion of the wild-type PAX3 protein have also identified a transcriptional inhibitory domain (Fig. 3) (Chalepakis et al. 1994b). This domain was localized to the NH_2-terminal 90 amino acid residues of the protein which includes the NH_2-terminal portion of the paired domain. In addition, a second, less potent inhibitory domain was localized to the vicinity of the homeodomain. Competition between these inhibitory domains and the COOH-terminal activation domain may explain the narrow concentration range over which PAX3 demonstrates transcriptional stimulation. In addition, the finding of similar transcriptional potencies of the COOH-terminal PAX3 and FKHR domains (Bennicelli et al. 1995) and contrasting potencies of the full-length PAX3 and PAX3-FKHR proteins (Fredericks et al. 1995) may be the result of differing interactions of the PAX3 NH_2-terminal inhibitory domain with the two structurally distinct COOH-terminal activation domains.

9 Phenotypic Consequences of Expression of PAX3 and PAX3-FKHR Proteins

In contrast to the loss of function associated with PAX3 mutations in Waardenburg syndrome and *splotch* (CHALEPAKIS et al. 1994a; UNDERHILL et al. 1995), these transcriptional studies (FREDERICKS et al. 1995) indicate that the alteration of the *PAX3* gene by the t(2;13) translocation results in a gain of function. These studies are consistent with the hypothesis that the t(2;13) translocation activates the oncogenic potential of PAX3 by dysregulating or exaggerating its normal function in the myogenic lineage. Clues for this normal function may be deduced from the skeletal muscle phenotype of homozygously mutated *splotch* mice (FRANZ et al. 1993; BOBER et al. 1994; GOULDING et al. 1994; WILLIAMS and ORDAHL 1994). In these animals, the limb musculature fails to develop whereas the axial musculature is reduced but develops relatively normally. The defect in limb musculature development appears to result from the failure of myogenic precursors to migrate from the somites into the limb buds. These findings suggest a role for PAX3 in the generation and/or migration of limb myogenic precursors. Possible functions of PAX3 thus include maintenance of a viable population by stimulating growth or inhibiting apoptosis, inhibition of premature myogenic differentiation, or direct stimulation of motility and dissemination. Exaggeration of these activities by the PAX3-FKHR fusion suggests several hypotheses relevant to the pathogenesis of ARMS, a highly aggressive tumor associated with the myogenic lineage.

Gene transfer studies in several model systems have investigated specific functions of PAX3 and PAX3-FKHR. In one study, stable transfection of PAX3 cDNA or several other paired box genes into NIH 3T3 or Rat-1 cells resulted in cellular transformation as indicated by focus formation in confluent monolayers, colony formation in soft agar, and tumor formation in nude mice (MAULBECKER and GRUSS 1993). These findings suggest a role for PAX3 and other paired box family members in growth control. However, the interpretation of these findings is complicated by lack of reproducibility in other laboratories (unpublished data). Another study focused on the ability of these proteins to inhibit the myogenic differentiation of C2C12 myoblasts or MyoD-expressing 10T1/2 cells (EPSTEIN et al. 1995). Differentiation of these cell lines is typically induced by the withdrawal of growth factors and results in myosin expression in the majority of cells. Transfection of PAX3 cDNA reduced the percentage of myosin-expressing colonies from 86%-87% to 40%-43%, and PAX3-FKHR transfection further reduced this percentage to 17%-27%. Therefore, these studies suggest a role for PAX3 in the control of myogenic differentiation and exaggeration of this role by the PAX3-FKHR fusion protein.

10 Final Perspectives

In summary, the aggressive pediatric soft tissue cancer ARMS is associated with a characteristic t(2;13) or variant t(1;13) chromosomal translocation (Fig. 1). These translocations juxtapose the transcription factor-encoding genes *PAX3* or *PAX7* with *FKHR* to generate fusion products (Fig. 4). These fusion proteins contain the intact PAX3 or PAX7 DNA binding domain joined to a truncated FKHR DNA binding domain and intact COOH-terminal *trans*-activation domain. The PAX3-FKHR fusion protein is a functional transcription factor, which has been shown in certain settings to be a much more potent transcriptional activator than wild-type PAX3. The PAX3-FKHR and PAX7-FKHR fusion proteins are therefore hypothesized to lead to aberrant or inappropriate expression of a set of genes with binding sites which are normally targeted during development by the wild-type PAX3 or PAX7 proteins. The products of these target genes are hypothesized to function normally in the growth and development of the myogenic lineage. The aberrant or inappropriate expression of these gene products would thus contribute to oncogenic initiation or progression by stimulating cell behaviors such as growth and motility or inhibiting cell behaviors such as apoptosis and terminal differentiation. Future investigations will refine the alterations in gene expression associated with the translocations, identify target genes, and elucidate the normal function and oncogenic activity of the corresponding gene products. These experiments will ultimately expand the understanding of the use of this molecular biology data in diagnostic settings and will indicate directions in which possible therapeutic strategies can be designed to interrupt these pathways.

Acknowledgements. The author thanks Robert Wilson, Richard Davis, and Jeannette Bennicelli for their helpful discussions and advice.

References

Baldwin CT, Hoth CF, Amos JA, da-Silva EO, Milunsky A (1992) An exonic mutation in the HuP2 paired domain gene causes Waardenburg's syndrome. Nature 355: 637–638

Barr FG, Chatten J, D'Cruz CM, Wilson AE, Nauta LE, Nycum LM, Biegel JA, Womer RB (1995) Molecular assays for chromosomal translocations in the diagnosis of pediatric soft tissue sarcomas. JAMA 27: 553–557

Barr FG, Galili N, Holick J, Biegel JA, Rovera G, Emanuel BS (1993) Rearrangement of the PAX3 paired box gene in the paediatric solid tumour alveolar rhabdomyosarcoma. Nature Genet 3: 113–117

Barr FG, Holick J, Nycum L, Biegel JA, Emanuel BS (1992) Localization of the t(2;13) breakpoint of alveolar rhabdomyosarcoma on a physical map of chromosome 2. Genomics 13: 1150–1156

Barr FG, Sellinger B, Emanuel BS (1991) Localization of the rhabdomyosarcoma t(2;13) breakpoint on a physical map of chromosome 13. Genomics 11: 941–947

Bennicelli JL, Fredericks WJ, Wilson RB, Rauscher FJ III, Barr FG (1995) Wild type PAX3 protein and the PAX3-FKHR fusion protein of alveolar rhabdomyosarcoma contain potent, structurally distinct transcriptional activation domains. Oncogene 11: 119–130

Biegel JA, Meek RS, Parmiter AH, Conard K, Emanuel BS (1991) Chromosomal translocation t(1;13)(p36;q14) in a case of rhabdomyosarcoma. Genes Chromosom Caner 3: 483–484
Bober E, Franz T, Arnold HH, Gruss P, Tremblay P (1994) *Pax-3* is required for the development of limb muscles: a possible role for the migration of dermomyotomal muscle progenitor cells. Development 120: 603–612
Bowcock AM, Farrer LA, Hebert JM, Bale AE, Cavalli-Sforza LL (1991) A contiguous linkage map for chromosome 13q with 39 distinct loci separated on average by 5.1 centimorgans. Genomics 11: 517–529
Burri M, Tromvoukis Y, Bopp D, Frigerio G, Noll M (1989) Conservation of the paired domain in metazoans and its structure in three isolated human genes. EMBO J 8: 1183–1190
Chalepakis G, Goulding M, Read A, Strachan T, Gruss P (1994a) Molecular basis of splotch and Waardenburg *Pax-3* mutations. Proc Natl Acad Sci USA 91: 3685–3689
Chalepakis G, Jones FS, Edelman GM, Gruss P (1994b) Pax-3 contains domains for transcription activation and transcription inhibition. Proc Natl Acad Sci USA 91: 12745–12749
Clark KL, Halay ED, Lai E, Burley SK (1993) Co-crystal structure of the HNF-3/*fork head* DNA-recognition motif resembles histone H5. Nature 364: 412–420
Clevidence DE, Overdier DG, Tao W, Qian X, Pani L, Lai E, Costa RH (1993) Identification of nine tissue-specific transcription factors of the hepatocyte nuclear factor 3/forkhead DNA binding-domain family. Proc Natl Acad Sci USA 90: 3948–3952
Czerny T, Schaffner G, Busslinger M (1993) DNA sequence recognition by Pax proteins: bipartite structure of the paired domain and its binding site. Genes Dev 7: 2048–2061
Davis RJ, D'Cruz CM, Lovell MA, Biegel JA, Barr FG (1994) Fusion of PAX7 to FKHR by the variant t(1;13)(p36;q14) translocation in alveolar rhabdomyosarcoma. Cancer Res 54: 2869–2872
Douglass EC, Rowe ST, Valentine M, Parham DM, Berkow R, Bowman WP, Maurer HM (1991) Variant translocations of chromosome 13 in alveolar rhabdomyosarcoma. Genes Chromosom Cancer 3: 480–482
Douglass EC, Valentine M, Etcubanas E, Parham D, Webber BL, Houghton PJ, Green AA (1987) A specific chromosomal abnormality in rhabdomyosarcoma. Cytogenet Cell Genet 45: 148–155
Downing JR, Khandekar A, Shurtlett SA, Head DR, Parham DM, Webber BL, Pappo AS, Hulshof MG, Conn WP, Shapiro DN (1995) Multiplex RT-PCR assay for the differential diagnosis of alveolar rhabdomyosarcoma and Ewing's sarcoma. Amer J Pathol 146: 626–634
Epstein JA, Lam P, Jepeal L, Maas RM, Shapiro DN (1995) Pax3 inhibits myogenic differentiation of cultured myoblast cells. J Biol Chem 270: 11719–11722
Epstein JA, Cai J, Glaser T, Jepeal L, Maas RM (1994a) Identification of a *Pax* paired domain recognition sequence and evidence for DNA-dependent conformational changes. J Biol Chem 269: 8355–8361
Epstein JA, Cai J, Maas RM (1994b) Pax3 recognizes a sequence within the 3′ UTR of the murine neurofibromatosis gene Nf1. Circulation 90: I634
Epstein DJ, Vekemans M, Gros P (1991) *splotch* (Sp^{2H}), a mutation affecting development of the mouse neural tube, shows a deletion within the paired homeodomain of *Pax-3*. Cell 67: 767–774
Franz T, Kothary R, Surani MAH, Halata Z, Grim M (1993) The Splotch mutation interferes with muscle development in the limbs. Anat Embryol 187: 153–160
Fredericks WJ, Galili N, Mukhopadhyay S, Rovera G, Bennicelli J, Barr FG, Rauscher FJ III (1995) The PAX3-FKHR fusion protein created by the t(2;13) translocation in alveolar rhabdomyosarcoma is a more potent transcriptional activator than PAX3. Mol Cell Biol 15: 1522–35
Galili N, Davis RJ, Fredericks WJ, Mukhopadhyay S, Rauscher FJ III, Emanuel BS, Rovera G, Barr FG (1993) Fusion of a fork head domain gene to PAX3 in the solid tumor alveolar rhabdomyosarcoma. Nature Genet 5: 230–235
Goulding M, Lumsden A, Paquette AJ (1994) Regulation of *Pax-3*, expression in the dermomyotome and its role in muscle development. Development 120: 957–971
Goulding MD, Chalepakis G, Deutsch U, Erselius JR, Gruss P (1991) Pax-3, a novel murine DNA binding protein expressed during early neurogenesis. EMBO J 10: 1135–1147
Han K, Manley JL (1993) Transcription repression by the *Drosophila* Even-skipped protein: definition of a minimal repression domain. Genes Dev 7: 491–503
Heim S, Mitelman F (1995) Cancer cytogenetics, 2nd edn. Wiley-Liss, New York
Jostes B, Walther C, Gruss P (1991) The murine paired box gene, Pax7, is expressed specifically during the development of the nervous and muscular system. Mech Dev 33: 27–37
Koi M, Johnson LA, Kalikin LM, Little PFR, Nakamura Y, Feinberg AP (1993) Tumor cell growth arrest caused by subchromosomal transferable DNA fragments from chromosome 11. Science 260: 361–364

Lai E, Clark KL, Burley SK, Darnell JE (1993) Hepatocyte nuclear factor 3/fork head or "winged helix" proteins: A family of transcription factors of diverse biologic function. Proc Natl Acad Sci USA 90: 10421–10423

Lai E, Prezioso VR, Smith E, Litvin O, Costa RH, Darnell JE (1990) HNF-3A, a hepatocyte enriched transcription factor of novel structure is regulated transcriptionally. Genes Dev 4: 1427–1436

Li J, Vogt PK (1993) The retroviral oncogene qin belongs to the transcription factor family that includes the homeotic gene fork head. Proc Natl Acad Sci USA 90: 4490–4494

Malo D, Schurr E, Epstein DJ, Vekemans M, Skamene E, Gros P (1991) The host resistance locus *Bcg* is tightly linked to a group of cytoskeleton-associated protein genes that include villin and desmin. Genomics 10: 356–364

Maulbecker CC, Gruss P (1993) The oncogenic potential of Pax genes. EMBO J 12: 2361–2367

Mitchell CD, Ventris JA, Warr TJ, Cowell JK (1991) Molecular definition in a somatic cell hybrid of a specific 2:13 translocation breakpoint in childhood rhabdomyosarcoma. Oncogene 6: 89–92

Nehls M, Pfeifer D, Schorpp M, Hedrich H, Boehm T (1994) New member of the winged-helix protein family disrupted in mouse and rate nude mutations. Nature 372: 103–107

O'Connell P, Lathrop GM, Nakamura Y, Leppert ML, Lalouel JM, White R (1989) Twenty loci form a continuous linkage map of markers for human chromosome 2. Genomics 5: 738–745

Parry P, Wei Y, Evans G (1994) Cloning and characterization of the t(X;11) breakpoint from a leukemic cell line identify a new member of the forkhead gene family. Genes Chromosom Cancer 11: 79–84

Raney RB, Hays DM, Tefft M, Triche TJ (1993) Rhabdomyosarcoma and the undifferentiated sarcomas. In: Pizzo PA, Poplack PG (eds) Principles and practice of pediatric oncology. JB Lippincott, Philadelphia, pp 769–794

Ren R, Mayer BJ, Cicchetti P, Baltimore D (1993) Identification of a ten-amino acid proline-rich SH3 binding site. Science 259: 1157–1161

Sadowski I, Ma J, Triezenberg S, Ptashne M (1988) GAL4-VP16 is an unusually potent transcriptional activator. Nature 335: 563–564

Sawyer JR, Crussi FG, Kletzel M (1994) Pericentric inversion (2)(p15q35) in an alveolar rhabdomyosarcoma. Cancer Genet Cytogenet 78: 214–218

Schäfer BW, Czerny T, Bernasconi M, Genini M, Busslinger M (1994) Molecular cloning and characterization of a human PAX-7 cDNA expressed in normal and neoplastic myocytes. Nucleic Acids Res 22: 4574–4582

Schäfer BW, Mattei MG (1993) The human paired domain gene PAX7 (Hup1) maps to chromosome 1p35-1p36.2. Genomics 17: 249–251

Scrable H, Witte D, Shimada H, Seemayer T, Wang-Wuu S, Soukup S, Koufos A, Houghton P, Lampkin B, Cavenee W (1989) Molecular differential pathology of rhabdomyosarcoma. Genes Chromosom Cancer 1: 23–25

Shapiro DN, Valentine MB, Sublett JE, Sinclair AE, Tereba AM, Scheffer H, Buys CHCM, Look AT (1992) Chromosomal sublocalization of the 2;13 translocation breakpoint in alveolar rhabdomyosarcoma. Genes Chromosom Cancer 4: 241–249

Shapiro DN, Sublett JE, Li B, Downing JR, Naeve CW (1993a) Fusion of PAX3 to a member of the forkhead family of transcription factors in human alveolar rhabdomyosarcoma. Cancer Res 53: 5108–5112

Shapiro DN, Sublett JE, Li B, Valentine MB, Morris SW, Noll M (1993b) The gene for PAX7, a member of the paired-box-containing genes, is localized on human chromosome arm 1p36. Genomics 17: 767–769

Stapleton P, Weith A, Urbanek P, Kozmik Z, Busslinger M (1993) Chromosomal localization of seven PAX genes and cloning of a novel family member, PAX-9. Nature Genet 3: 292–298

Strachan T, Read AP (1994) PAX genes. Curr Op Gen Dev 4: 427–438

Tassabehji M, Read AP, Newton VE, Harris R, Ballings R, Gruss P, Strachan T (1992) Waardenburg's syndrome patients have mutations in the human homologue of the Pax-3 paired box gene. Nature 355: 635–636

Treisman J, Harris E, Desplan C (1991) The paired box encodes a second DNA-binding domain in the paired homeodomain protein. Genes Dev 5: 594–604

Tremblay P, Gruss P (1994) PAX: Genes for mice and men. Pharmac Ther 61: 205–226

Tsokos M (1994) The diagnosis and classification of childhood rhabdomyosarcoma. Sem Diag Path 11: 26–38

Tsokos M, Webber BL, Parham DM, Wesley RA, Miser A, Miser JS, Etcubanas E, Kinsella T, Grayson J, Glatstein E, Pizzo PA, Triche TJ (1992) Rhabdomyosarcoma – a new classification scheme related to prognosis. Arch Pathol Lab Med 116: 847–855

Turc-Carel C, Lizard-Nacol S, Justrabo E, Favrot M, Philip T, Tabone E (1986) Consistent chromosomal translocation in alveolar rhabdomyosarcoma. Cancer Genet Cytogenet 19: 361–362
Underhill DA, Vogan KJ, Gros P (1995) Analysis of the mouse Splotch-delayed mutation indicates that the PAX-3 paired domain can influence homeodomain DNA-binding activity. Proc Natl Acad Sci USA 92: 3692–3696
Valentine M, Douglass EC, Look AT (1989) Closely linked loci on the long arm of chromosome 13 flank a specific 2;13 translocation breakpoint in childhood rhabdomyosarcoma. Cytogenet Cell Genet 52: 128–132
Wang-Wuu S, Soukup S, Ballard E, Gotwals B, Lampkin B (1988) Chromosomal analysis of sixteen human rhabdomyosarcomas. Cancer Res 48: 983–987
Watson ML, D'Eustachio P, Mock BA, Steinberg AD, Morse HC, Oakey RJ, Howard TA, Rochelle JM, Seldin MF (1992) A linkage map of mouse chromosome 1 using an interspecific cross segregating for the *gld* autoimmunity mutation. Mammalian Genome 2: 158–171
Weigel D, Jackle H (1990) The fork head domain: A novel DNA binding motif of eukaryotic transcription factors? Cell 63: 455-456
Weigel D, Jurgens G, Kuttner F, Seifert E, Jackle H (1989) The homeotic gene *fork head* encodes a nuclear protein and is expressed in the terminal regions of the Drosophila embryo. Cell 57: 645–658
Whang-Peng J, Knutsen T, Theil K, Horowitz ME, Triche T (1992) Cytogenetic studies in subgroups of rhabdomyosarcoma. Genes Chromosom Cancer 5: 299–310
Whang-Peng J, Triche TJ, Knutsen T, Miser J, Kao-Shan S, Tsai S, Israel MA (1986) Cytogenetic characterization of selected small round cell tumors of childhood. Cancer Genet Cytogenet 21: 185–208
Williams BA, Ordahl CP (1994) Pax-3 expression in segmental mesoderm marks early stages in myogenic cell specification. Development 120: 785–796

TLS-CHOP and the Role of RNA-Binding Proteins in Oncogenic Transformation

D. Ron

1 Introduction

Many of the genes found to be altered in cancer cells encode proteins with regulatory effects on the pattern of gene expression. Some, like growth factor receptors and adapter molecules, do so indirectly by interfering with signaling pathways that eventually converge on a nuclear target. Others are more directly associated with regulation of gene expressions, being either DNA-binding transcription factors or proximate regulators of transcription factor function.

The association of transcription factors with oncogenesis is consistent with the idea that the complement of genes expressed determines cellular phenotype. A clue as to the magnitude of the contribution of altered gene expression to the process of oncogenesis is provided by the observation that such regulatory proteins are found to be encoded by many virally transduced oncogenes (Varmus 1984). However, in human cancers a viral etiology seems less important than in other organisms. In our species tumor cytogenetics presents the most compelling epidemiological argument in favor of a widespread role for transcription factors and signaling molecules as effectors of the neoplastic phenotype (Bishop 1987). In a proportion of cancer cells the clonal abnormality is associated with rearrangement of the genetic material that leads to a recognizable alteration in chromosomal structure. Originally pursued by researchers as ultrastructural markers of the abnormal clone, recent years have

Skirball Institute of Biomolecular Medicine, Departments of Medicine and Cell Biology and the Kaplan Cancer Center, New York University Medical Center, New York, NY 10016, USA

witnessed the molecular characterization of some of the involved genes and, as this volume indicates, many of them turned out to encode transcription factors and signaling molecules (RABBITTS 1994).

Here I will review our understanding of the molecular consequences of the chromosomal rearrangement common to human myxoid liposarcoma – a tumor of adipose tissue. What makes the myxoid liposarcoma oncogene especially interesting is the fact that a transcription factor, CHOP, is fused to a discrete domain of a novel RNA-binding protein in an arrangement that seems common to other sarcomas.

2 CHOP and C/EBP Isoforms in Adipose Tissue

CHOP (also known as GADD153) is a small nuclear protein whose identification was based on its ability to dimerize with members of the C/EBP family of transcription factors (RON and HABENER 1992). This family of regulatory molecules binds to and activates the promoters of multiple genes including many involved in intermediary metabolism, cytokine cascades and cellular differentiation events (reviewed in MCKNIGHT et al. 1989). Target sequence recognition by C/EBPs is strictly dependent on protein dimerization and both events are mediated by a C-terminus basic region leucine-zipper (bZIP) domain (LANDSCHULZ et al. 1989). The extreme degree of amino acid sequence conservation at the bZIP domain appears to explain why the C/EBP family members (known as C/EBP isoforms) heterodimerize promiscuously and bind indistinguishable target sequences (WILLIAMS et al. 1991). In the adipose lineage a finely coordinated cascade of induced expression of C/EBP isoforms plays an important role in promoting differentiation (CAO et al. 1991).

CHOP does not bind classic C/EBP sites. This striking deviation from the properties common to all other known C/EBP isoforms is explained by the unusual structure of the CHOP basic region (RON and HABENER 1992). Incapable of homodimerization, CHOP forms stable heterodimers with C/EBPα and β and these dimers are directed away from classic C/EBP sites (RON and HABENER 1992; UBEDA et al. 1995). From the perspective of classic C/EBP sites CHOP therefore functions as a dominant negative inhibitor of C/EBP activation (RON and HABENER 1992; UBEDA et al. 1995).

In contrast with its C/EBP dimerization partners that are abundant proteins present in many different cell types, CHOP is expressed at very low levels under normal conditions (FORNACE et al. 1989; RON and HABENER 1992). However, a variety of stressful events rapidly induce the transcription of the *chop* gene resulting in the accumulation of the protein in the nucleus. This explains why *chop* has also been isolated in a screen that looked for genes induced by DNA-damage (and given the name *gadd153*, FORNACE et al. 1989). Though *chop* can be induced by some genotoxic agents an additional major pathway for its induction appears to entail various forms of metabolic stress. Culture in low glucose (Carlson et al. 1993), amino acid depletion (Marten et al. 1994), hypobaric oxygen levels and a variety of agents that interfere with proper protein synthesis and folding (Price and Calderwood 1992) all induce *chop* expression.

Functional analyses of the consequence of *chop* expression support the idea that the protein plays a role as an effector of the cellular adaptation to stress. Expression of high level of wild-type CHOP protein in dividing cells causes cell cycle arrest at the G1/S boundary (Barone et al. 1994; Zhan et al. 1994). This growth suppressing activity of CHOP requires both an intact leucine-zipper dimerization domain and the adjacent basic region. The requirement for an intact basic region is most consistent with the existence of target genes capable of responding to the CHOP-C/EBP signal and mediating the effects on cellular growth. The implication of CHOP in a growth suppressing signaling pathway is easy to rationalize in terms of the need for cells to respond to unfavorable culture conditions with growth arrest (reviewed in Ron 1994).

In pre-adipocytes the ectopic expression of CHOP leads to an attenuation of differentiation to mature adipose cells (Batchvarova et al. 1995). The inhibitory effect of stress-induced CHOP can here be understood in terms of the cell's need to suppress the metabolically expensive process of differentiation when confronted with untoward culture conditions. The decisive role C/EBP proteins play in adipocytic differentiation (Freytag et al. 1994; Lin and Lane 1992, 1994; Samuelsson et al. 1991) and the demonstrated ability of CHOP to inhibit their activity suggest that inhibition of C/EBP protein function may be important to the anti-adipogenic activity of CHOP. In addition to a purely inhibitory role, CHOP-C/EBP heterodimers may also be active in regulating the expression of specific target genes. This latter possibility is supported by the observation that CHOP inhibits adipogenesis at substoichiometric concentrations, and the DNA-contacting basic region of CHOP is required for the inhibitory effect (Batchvarova et al. 1995). While CHOP target genes have yet to be identified, in vitro work with bacterially expressed CHOP protein has led to the recognition of specific DNA sequences capable of serving as C/EBP-CHOP binding sites (Ubeda et al. 1995).

3 TLS-CHOP and Myxoid Liposarcoma

Human *CHOP* is located on chromosome 12q13.1 (Park et al. 1992), the site of a breakpoint frequently found in myxoid liposarcoma (MLPS, Sreekantiaiah et al. 1992; Turc-Carel et al. 1986). In these tumors a reciprocal balanced translocation, t(12;16)(q13;p11), juxtaposes DNA sequence from the short arm of chromosome 16p at band 11 to the long arm of chromosome 12q at band 13. The reciprocal derivative chromosome 16 can be lost in some tumor cases and hence is unlikely to contribute to the oncogenic event.

Southern blot analysis of DNA from patients with MLPS revealed that in virtually all cases the CHOP gene is structurally rearranged in the tumors (Åman et al. 1992). This structural rearrangement is associated with the constitutive presence of an abnormally large CHOP mRNA in tumor cells. Cloning of the abnormal-sized constitutive CHOP cDNA from MLPS cells revealed it to consist of the full-length CHOP coding region fused in frame 3′ of a novel gene that we termed *TLS* (Fig. 1A, Crozat et al. 1993; also named *FUS*, Rabbitts et al. 1993). The *TLS* regulatory sequences drive the expression of this fusion gene. Because *TLS*, in contrast to *CHOP* is constitutively expressed in adipose tissue, the *TLS-CHOP* fusion gene is also constitutively expressed in the tumor cells.

TLS-CHOP encodes an abundant nuclear protein. In MLPS cells this protein exists predominantly as a heterodimer with C/EBPβ. This suggests that if it plays a role in tumor formation it is likely to do so in the context of dimerization with other C/EBP family members (Barone et al. 1994; Zinszner et al. 1994). To determine if TLS-CHOP transforms cells by a dominant mechanism we constructed recombinant retroviruses that encode the oncoprotein and tested their ability to induce cellular transformation in NIH 3T3 cells. TLS-CHOP led to the acquisition of the ability to grow as colonies in soft agar, form foci on confluent monolayers and give rise to tumors when injected into nude mice (Zinszner et al. 1994). Deletion of the TLS N-terminal sequence, the DNA-binding basic region or the leucine-zipper dimerization domain of CHOP all led to loss of transforming activity. These results fit nicely with the observation that presence of TLS sequence abolishes the growth arresting properties of CHOP in cellular microinjection experiments (Barone et al. 1994).

4 Mechanism of Action of the TLS-CHOP Oncogene

4.1 Transcriptional Activation

TLS is approximately 55% identical throughout its length to another human gene product EWS (Fig. 1B). The latter is implicated in a variety of tumor specific translocation events that fuse its N-terminal to the C-terminal DNA binding domain of one of several transcription factors (Fig. 1C references: Delattre et al. 1992; Ladanyi and Gerald 1994; Zucman et al. 1993a; Zucman et al. 1993b, also see related reviews in this volume). This similarity in architecture between the TLS-derived and EWS-derived fusion oncogenes and the similarity in structure between TLS and EWS led others and us to examine the N-terminal domain of these two proteins more closely. Both domains were found to have strong transcriptional activation potential when bound to a reporter gene's promoter (Bailly et al. 1994; May et al. 1993a,b; Prasad et al. 1994; Zinszner et al. 1994). This result is easy to rationalize in terms of a model whereby the fusion oncogene transforms cells because it is a stronger activator of effector target genes with which the DNA-binding

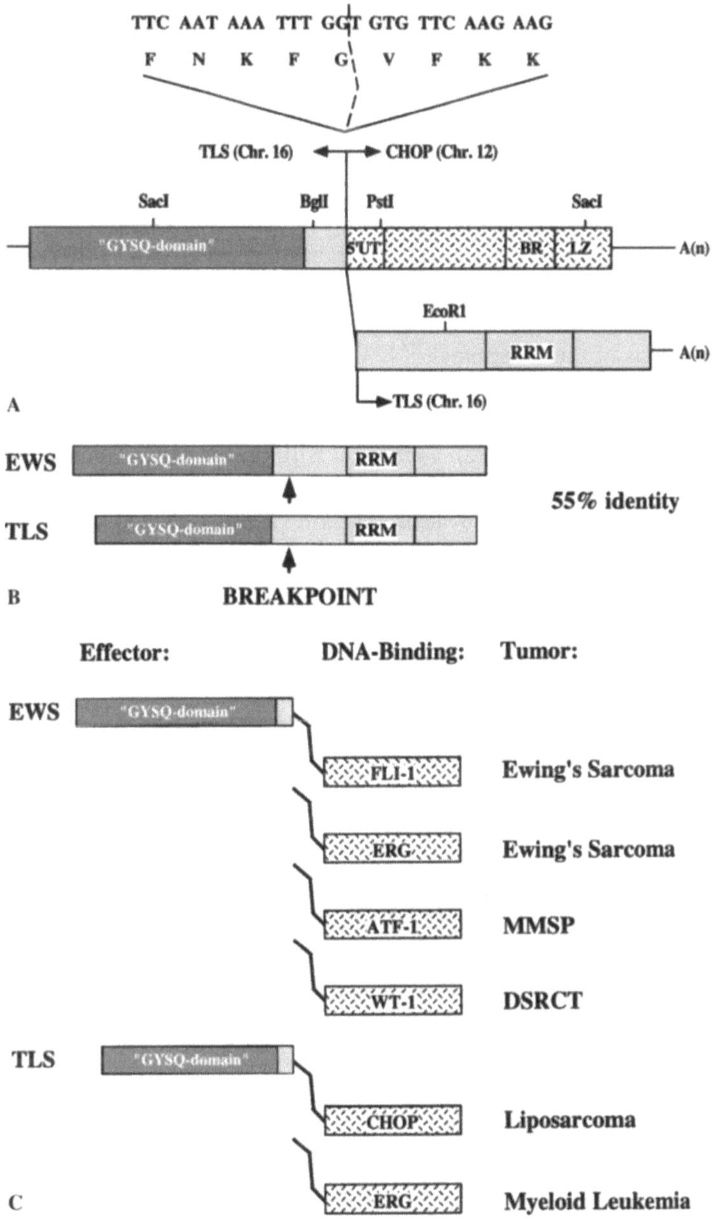

Fig. 1A-C. TLS and EWS participate in the formation of similarly structured oncogenic fusion proteins. **A** Structure of the TLS-CHOP fusion mRNA, the germline TLS mRNA and features of the encoded proteins. The sequence of the fusion junction is shown *above*. The GYSQ-rich domain, the 5́ normally untranslated region of CHOP (5'UT), the DNA-binding CHOP basic region (BR), the CHOP leucine-zipper dimerization domain (LZ) and the TLS RNA recognition motif (RRM) are all indicated. **B** Similarity in architecture between TLS and EWS. The *arrow* points to the commonly occurring site of gene fusion in the various sarcomas. **C** Comparison of structure of EWS and TLS fusion oncoproteins. MMSP refers to malignant melanoma of the soft part and DSRCT refers to diffuse small round cell tumors. References for the original descriptions of these fusion oncogenes are provided in the body of the text

component interacts. Such activation would also benefit from the high transcriptional activity of the *TLS* and *EWS* promoter that leads to constitutively high levels of the respective fusion oncoprotein. This mechanism may indeed be very important in the case of EWS-FLI1 (the Ewing's sarcoma oncogene), in which activation by EWS-FLI1 far exceeds that of normal FLI1 (Bailly et al. 1994; Mao et al. 1994; May et al. 1993b). However, there may be additional requirements of TLS/EWS that suggest specificity for a distinct type of activation domain. At least two powerful heterologous transcriptional activation domains from C/EBPα and HSV-VP16 proteins do not fully substitute for the TLS N-terminus in converting CHOP to an oncogene (Zinszner et al. 1994).

Deletional analysis of the N-terminal of EWS suggests that the transcriptional activation potential is partially linked to total content of N-terminal sequence – progressive deletion of the N-terminal leads to progressive decrease in transcriptional activation (Lessnick et al. 1995). Consistent with this notion of "activation by content", it is noteworthy that the N-terminals of TLS and EWS contain multiple repeats of six to nine amino acids that are rich in G,Y,S,Q with only minimal degree of local positional preference. Interestingly, a similar aminoacid composition is also found in another fusion oncoprotein, SYT, involved in synovial sarcomas with the t(X;18)(p11.2;q11.2) translocation (Clark et al. 1994). That this type of domain would be found in three independently identified sarcoma fusion oncogenes suggests specificity for oncogenesis in soft tissues.

The striking similarities between these sacroma-associated GYSQ domains even raises the possibility that the DNA-binding component of the fusion oncogenes plays merely an ancillary role in the process of cellular transformation. To address this issue experimentally we have compared the histological appearance of tumors derived from NIH-3T3 cells transformed with TLS-CHOP, EWS-CHOP, TLS-FLI1 and EWS-FLI1. Regardless of its fusion partner, CHOP imparted a spindle cell sarcoma phenotype on the tumors whereas FLI1 was associated with a round cell morphology (Zinszner et al. 1994, Fig. 5 therein). These results indicate that the DNA-binding component participates in defining tumor phenotype and is consistent with the idea that the CHOP-based and FLI1-based oncoproteins contact different target genes.

4.2 TLS Is an RNA-Binding Protein

Examination of the C-terminal portion of the germline-encoded TLS protein reveals it to contain a structural motif commonly found in RNA-binding proteins. This 86 amino acid RNA recognition motif (RRM) is strikingly conserved between TLS and EWS (>80% identity) and this conservation extends to residues in which both proteins deviate from the consensus defined by many other RRM-containing proteins (Crozat et al. 1993, references therein). The C-terminal portions of TLS and EWS are capable of binding tightly to mRNA in vitro (Crozat et al. 1993; Ohno et al. 1994). In vivo TLS shuttles from the nucleus to the cytosol along with other RNA-binding proteins in a large ribonucleic protein complex (RNP), and this shuttling is disrupted by inhibiting mRNA gene transcription with agents such as α-amanitin (ZINSZNER et al. 1994 and unpublished observations). These findings strongly support a role for RNA-binding by TLS under normal conditions. TLS-CHOP does not bind RNA and does not shuttle – consistent with a role for the C-terminal portion of the molecule in mediating RNA-binding (ZINSZNER et al. 1994).

TLS is a relatively abundant molecule, suggesting that the protein may interact with the mRNA products of many genes. While the target gene specificity of TLS has not been addressed in mammalian cells, in *Drosophila melanogaster* a protein homologous to TLS and EWS, Caz (STOLOW and HAYNES 1995), has been localized to chromatin regions of most genes actively transcribed by RNA polymerase II (IMMANUEL et al. 1995). A similar distribution is exhibited by some hnRNPs (MATUNIS et al. 1992). This result suggests a role for TLS in some general aspect of pre-mRNA/mRNA metabolism.

4.3 Models

What might be the role of a transcriptional activation domain in the function of an RNA-binding protein? One possibility is that TLS serves to activate promoters using the nascent pre-mRNA as a scaffold (Fig. 2A). A precedent for this is set by TAT, an HIV-encoded RNA-binding protein that activates the HIV promoter from its TAR binding site on the nascent viral transcript (BERKHOUT et al. 1989; SELBY and PETERLIN 1990). It is also possible, however, that the transcriptional activation function of TLS (and EWS), reflected in the activity of the fusion oncogenes, has nothing to do with the role of the germline-encoded proteins. Transcriptional activation could be fortuitously unmasked by the gene fusion event. The observation that sequences that can function as activation regions are readily encoded in random bits of *E. coli* genomic DNA suggests that no very elaborate structure is required to display trans-activation potential and that such latent activity may be present in protein domains that normally have other functional roles (MA and PTASHNE 1987).

Recent studies have demonstrated that transcriptional activation can be achieved in many instances by a stable interaction between a DNA-bound factor and one of several different components of the basal transcriptional machinery (BARBERIS et al. 1995; CHATTERJEE and STRUHL 1995; KLAGES and STRUBIN 1995). This suggests that the interaction of the N-terminus of TLS with a component of the transcriptional machinery might have different consequences when the TLS N-terminal is presented in the context of an RNA-binding protein or a DNA-binding protein. An interaction of the N-terminal of TLS with the transcriptional machinery may serve to recruit this RNA-binding protein to the nascent transcript in the normal course of events (Fig. 2B). However, in the oncogenic fusion protein the same interaction may serve to recruit the transcriptional machinery to CHOP target genes – a reflection of the trans-activation potential of the N-terminal of TLS (Fig. 2A).

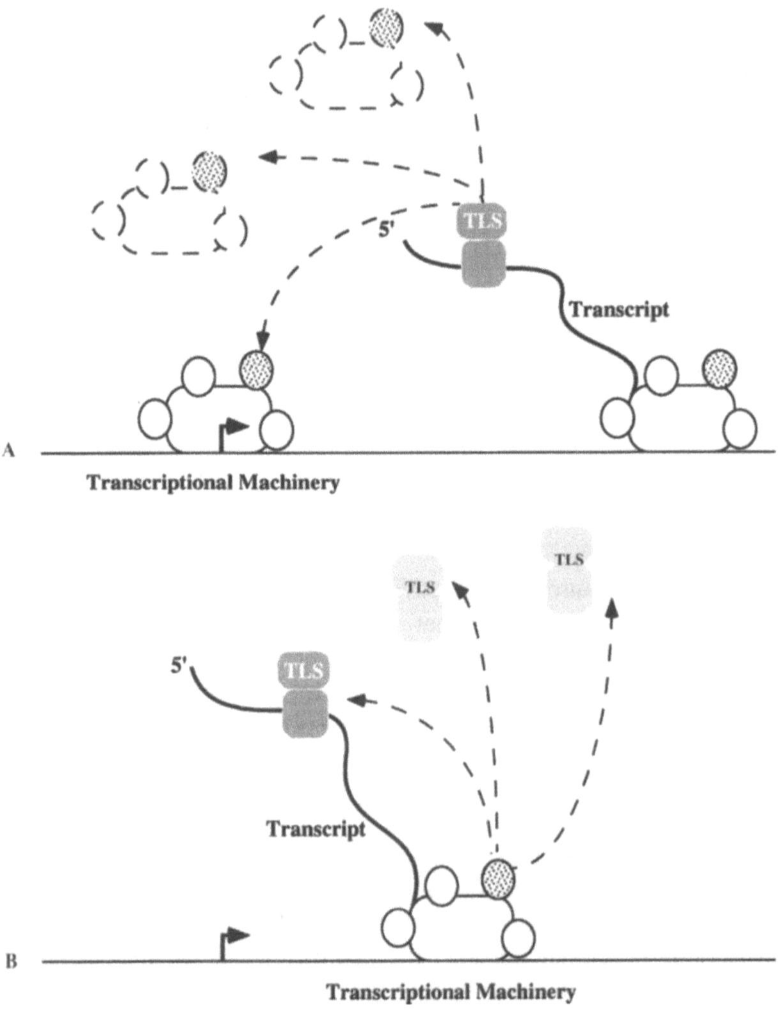

Fig. 2A, B. Two hypothetical roles for the transcriptional activation domain located in the N-terminus of TLS. **A** Interaction of RNA-bound TLS with a component(s) of the basic transcriptional machinery (*stippled circle*) serves to enhance recruitment of the RNA polymerase to the adjacent promoter. In this model TLS serves as a "transcription factor" that uses the nascent RNA transcript as a scaffold. **B** Interaction of the N-terminal of TLS with a component of the transcriptional machinery serves to recruit the RNA-binding protein to the nascent transcript. In this model it is the transcriptional machinery that directs TLS to its RNA target

How might the discrete activities of the individual domains of TLS-CHOP account for its transforming properties? Ectopic expression of CHOP has contradictory effects on cell growth. High level expression by microinjection of the protein or by high copy expression plasmid leads to cell cycle arrest at G1/S (Barone et al. 1994; Zhan et al. 1994), whereas more controlled levels of expression by retroviral transduction leads to increased rates of growth (though not quite frank transformation, ZINSZNER et al. 1994). It is possible that CHOP target genes participate in both growth arrest and transformation, with subtle differences in profile of expression having profound effects on cellular responses. TLS-CHOP may tip the balance in favor of transformation by altering the pattern of expression of such target genes. Contributors to deregulated expression of CHOP target genes may include the fact that TLS-CHOP is constitutively present whereas CHOP is only transiently induced by stress, and the presence of the TLS effector domain in TLS-CHOP which may send a different signal to the transcriptional machinery from that sent by CHOP alone (Fig. 3A).

TLS-CHOP exists in the cell as a stable dimer with C/EBPβ. The latter is a substrate for multiple kinases and appears to be positioned downstream of several signal transduction pathways (METZ and ZIFF, 1991; POLI et al. 1990; WEGNER et al. 1991). The TLS N-terminus may alter, in *cis*, some aspect of the interaction between C/EBPβ and its upstream regulators. The expression of C/EBPβ target genes could then be modified, with important consequences for cellular growth (BUCK

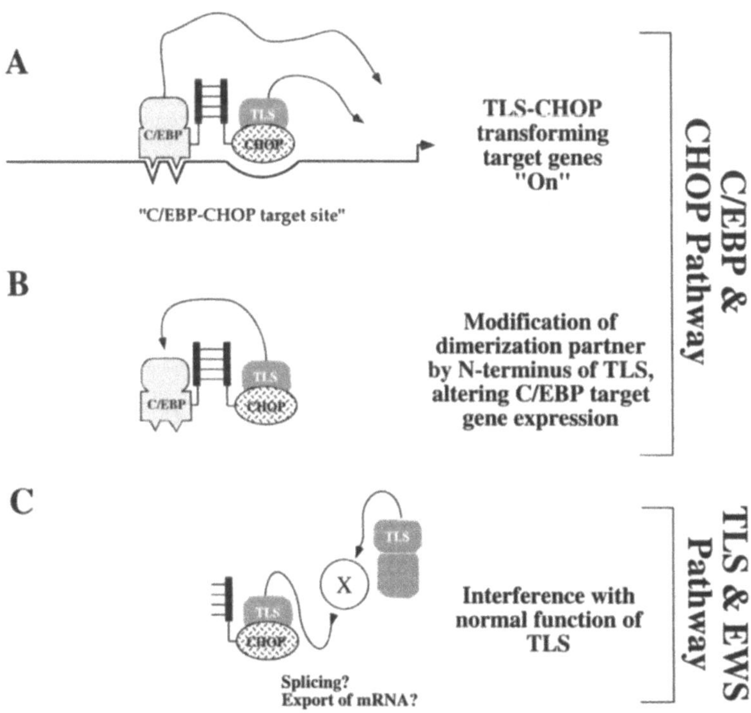

Fig. 3A-C. Three hypothetical components of the transformation process by the TLS-CHOP oncoprotein. **A** The TLS-CHOP-C/EBP heterodimer alters the expression of oncogenic downstream target genes by binding to sites in their promoters. **B** The TLS N-terminus acts to modify some aspect of the function of the C/EBP dimerization partner. This is translated to effects on gene expression via C/EBP target genes. **C** TLS-CHOP interferes with the function of germline-encoded TLS, for example, by competing for an essential target, "protein X"

et al. 1994). Such a "catalytic" mode of action for TLS-CHOP would not be mediated through CHOP target genes (Fig. 3B).

The normal function of TLS is not known. However, as an RNA-binding protein that shuttles from the nucleus to the cytosol TLS may regulate various aspects of gene expression posttranscriptionally (splicing, mRNA export, translation). TLS-CHOP may influence some critical aspect of normal TLS function, perhaps by a "dominant negative" mechanism of action, leading to a post-transcriptional alteration in regulated gene expression (Fig. 3C). Addressing this possibility will require better understanding of the normal role of TLS and identification of its cellular contingents. This component of the model is nonetheless appealing because it may be common to both TLS and EWS-based fusion oncogenes and may help explain the special link between sarcomas and these type of RNA-binding proteins. Finally, it is important to point out that the three models proposed here are not mutually exclusive – by simultaneously targeting different regulatory pathways TLS-CHOP and similar oncogenes may be multivalent and therefore unusually potent.

Acknowledgements. Thanks are due to my coworkers: Helene Zinszner, David Immanuel and Ricard Albalat for their contributions to the experiments described herein and to our collaborators in the lab of Felix Mittelman in Lund, Sweden for sharing reagents and thoughts. Naoko Tanese offered many insightful comments that have been incorporated into this manuscript.

Study of the pathogenesis of myxoid liposarcoma is supported by a grant from the USPHS (CA 60945). David Ron is a Pew Scholar in the Biomedical Sciences.

References

Åman P, Ron D, Mandahl N, Fioretos T, Heim S, Arhenden K, Willén H, Rydholm A, Mitelman F (1992) Rearrangement of the transcription factor gene *CHOP* in myxoid liposarcomas with t(12; 160(q13;p11)). Genes Chrom Cancer 5: 271–277

Bailly RA, Bousselut R, Zucman J, Cormier F, Delattre O, Roussel M, Thomas G, Ghysdeal J (1994) DNA-binding and transcriptional activation properties of the EWS-Fli1 fusion protein resulting from the t(11;22) translocation in Ewing sarcoma. Mol Cell Biol 14: 3230–3241

Barberis A, Pearlberg J, Simkovich N, Farrell S, Reinagel P, Bamdad C, Sigal G, Ptashne M (1995) Contact with a component of the polymerase II holoenzyme suffices for gene activation. Cell 81: 359–368

Barone MV, Crozat AY, Tabaee A, Philipson L, Ron D (1994) CHOP (GADD153) and its oncogenic variant, TLS-CHOP, differ in their ability to induce G1/S arrest. Genes Dev 8: 453–464

Batchvarova N, Wang X-Z, Ron D (1995) Inhibition of adipogenesis by the stress-induced protein CHOP (GADD153) EMBO J 14: 4654–4661

Berkhout B, Silverman RH, Jeang K-T (1989) Tat transactivates the human immunodeficiency virus through a nascent RNA target. Cell 59: 273–282

Bishop JM (1987) The molecular genetics of cancer. Science 235: 305–311

Buck M, Turler H, Chojkier M (1994) LAP (NF-IL-6), a tissue-specific transcriptional activator, is an inhibitor of hepatoma cell proliferation. EMBO J 13: 851–860

Cao Z, Umek RM, McKnight SL (1991) Regulated expression of three C/EBP isoforms during adipose conversion of 3T3-L1 cells. Genes Dev 5: 1538–1552

Carlson SG, Fawcett TW, Bartlett JD, Bernier M, Holbrook NJ (1993) Regulation of the C/EBP-related gene, *gadd*153, by glucose deprivation. Mol Cell Biol 13: 4736–4744

Chatterjee S, Struhl K (1995) Connecting a promoter-bound protein to TBP bypasses the need for a transcriptional activation domain. Nature 374: 820–822
Clark J, Rocques PJ, Crew AJ, Gill S, Shipley J, Chan AM-L, Gusterson BA, Cooper CS (1994) Identification of novel genes, SYT and SSX, involved in the t(X;18)(P11.2;q11.2) translocation found in human synovial sarcoma. Nat Gen 7: 502–508
Crozat AY, Åman P, Mandahl N, Ron D (1993) Fusion of CHOP to a novel RNA-binding protein in human myxoid liposarcoma with t(12;16)(q13;p11). Nature 363: 640–644
Delattre O, Zucman J, Plougastel B, Desmaze C, Melot T, Peter M, Kovar H, Joubert I, deJong P, Rouleau G, Aurias A, Thomas G (1992) Gene fusion with an ETS DNA-binding domain caused by chromosome translocation in human tumors. Nature 359: 162–165
Fornace AJ, Neibert DW, Hollander MC, Luethy JD, Papathanasiou M, Fragoli J, Holbrook NJ (1989) Mammalian genes coordinately regulated by growth arrest signals and DNA-damaging agents. Mol Cell Biol 9:4196–4203
Freytag SO, Paielli DL, Gilbert JD (1994) Ectopic expression of the CCAAT/enhancer-binding protein α promotes the adipogenic program in a variety of mouse fibroblastic cells. Genes Dev 8: 1654–1663
Immanuel D, Zinszner H, Ron D (1995) Association of SARFH (sarcoma associated RNA-binding fly homologue), with regions of chromatin transcribed by rNA polymerase II. Mol Cell Biol 15: 4562–4571
Klages N, Strubin M (1995) Stimulation of RNA polymerase II transcription initiation by recruitment of TBP in vivo. Nature 374: 822–823
Ladanyi M, Gerald W (1994) Fusion of the *EWS* and *WT1* genes in the desmoplastic small round cell tumor. Cancer Res 54: 2837–2840
Landschulz WH, Johnson PF, McKnight SL (1989) The DNA binding domain of the rat liver nuclear protein C/EBP is bipartite. Science 243: 1681–1688
Lessnick S, Braun B, Denny C, May W (1995) Multiple domains mediate transformation by the Ewing's sarcoma EWS/FLI-1 fusion gene. Oncogene 10: 423–431
Lin F-T, Lane M (1994) CCAAT/enhancer binding protein α is sufficient to initiate the 3T3-L1 adipocyte differentiation program. Proc Natl Acad Sci USA 91: 8757–8761
Lin F-T, Lane MD (1992) Antisense CCAAT/enhancer-binding protein RNA suppresses coordinate gene expression and triglyceride accumulation during differentiation of 3T3-L1 preadipocytes. Genes Dev 6: 533–544
Ma J, Ptashne M (1987) A new class of yeast transcriptional activators. Cell 51: 113–119
Mao X, Miesfeldt S, Yang H, Leiden J, Thompson C (1994) The FLI-1 and chimeric EWS-FLI-1 oncoproteins display similar DNA binding specificties. J Biol Chem 269: 18216–18222
Marten NW, Burke EJ, Hayden JM, Straus DS (1994) Effect of amino acid limitation on the expression of 19 genes in rat hepatoma cells. FASEB J 8: 538–544
Matunis EL, Matunis MJ, Dreyfuss G (1992) Drosophila heterogenous nuclear ribonucleoproteins. J Cell Biol 116: 257–269
May WA, Gishizky ML, Lessnick SL, Lunsford LB, Lewis BC, Delattre O, Zucman J, Thomas G, Denny CT (1993a) Ewing sarcoma 11;22 tranlocation produces a chimeric transcription factor that requires the DNA-binding domain encoded by FLI1 for transformation. Proc Natl Acad Sci USA 90: 5752–5756
May WA, Lessnick SL, Braun BS, Klemsz M, Lewis BC, Lunsford LB, Hromas R, Denny CT (1993b) The Ewing's sarcoma EWS/FLI-1 fusion gene encodes a more potent transcriptional activator and is a more powerful transforming gene than FLI-1. Mol Cell Biol 13: 7393–7398
McKnight SL, Lane MD, Gluecksohn-Waelsch S (1989) Is CCAAT/enhancer binding protein a central regulator of energy metabolism? Genes Dev 3: 2021–2024
Metz R, Ziff E (1991) cAMP stimulates the C/EBP-related transcription factor rNFIL-6 to *trans*-locate to the nucleus and induce *c-fos* transcription. Genes Dev 5: 1754–1766
Ohno T, Ouchida M, Lee L, Gatalica Z, Rao V, Reddy E (1994) The EWS gene, involved in Ewing family of tumors, malignant melanoma of soft parts and desmoplastic small round tumors, codes for an RNA binding proteins with novel regulatory domains. Oncogene 9: 3087–3097
Park JS, Luethy JD, Wang MG, Fragnolli J, Fornace AJ, McBride OW, Holbrook NJ (1992) Isolation, characterization and chromosomal localization of the human GADD153 gene. Gene 116: 259–267
Poli V, Mancini FP, Cortese R (1990) IL-6DBP, a nuclear protein involved in interleukin-6 signal transduction, defines a new family of leucine zipper proteins related to C/EBP. Cell 63: 643–653
Prasad D, Ouchida M, Lee L, VN R, Reddy E (1994) TLS/FUS fusion domain of TLS/FUS-erg chimeric protein resulting from the t(16;21) chromosomal translocation in human myeloid leukemia functions as a transcriptional activation domain. Oncogene 9: 3717–3729

Price B, Calderwood S (1992) Gadd 45 and Gadd 153 messenger RNA levels are increased during hypoxia and after exposure of cells to agents which elevate the levels of glucose-regulated proteins. Cancer Res 52: 3814–3817

Rabbitts TH (1994) Chromosomal translocations in human cancer. Nature 372: 143–149

Rabbits TH, Forster A, Larson R, Nathan P (1993) Fusion of the dominant negative transcription regulator CHOP with a novel gene FUS by translocation t(12;16) in malignant liposarcoma. Nat Gen 4: 175–180

Ron D (1994) Inducible growth arrest: New mechanistic insights. Proc Natl Acad Sci USA 91: 1985–1986

Ron D, Habener JF (1992) CHOP, a novel developmentally regulated nuclear protein that dimerizes with transcription factors C/EBP and LAP and functions as a dominant negative inhibitor of gene transcription. Genes Dev 6: 439–453

Samuelsson L, Strömberg K, Vikman K, Bjursell G, Enerbäck S (1991) The CCAAT/enhancer binding protein and its role in adipocyte differentiation: evidence for direct involvement in terminal adipocyte development. EMBO J 10: 3787–3793

Selby MJ, Peterlin BM (1990) Trans-activation by TAT via a heterologous RNA binding protein. Cell 62: 769–776

Sreekantiaiah S, Karakousis CP, Leong SPL, Sandberg AA (1992) Cytogenetic findings in liposarcoma correlate with histopathologic subtypes. Cancer 69: 2484–2495

Stolow D, Haynes S (1995) Cabeza, a Drosophila gene encoding a novel RNA binding protein, shares homology with EWS and TLS, two genes involved in human sarcoma formation. Nucleic Acids Res 23: 835–843

Turc-Carel C, Limon J, Dal Cin P, Rao U, Karakousis C, Sandberg AA (1986) Cytogenetic studies of adipose tissue tumors. II. Recurrent reciprocal translocation t(12;16)(q13;p11) in myxoid liposarcoma. Cancer Genet Cytogenet 23: 291–299

Ubeda M, Zinszner H, Wang X-Z, Wu I, Habener J, Ron D (1995) Stress-induced binding of transcription factor CHOP to a novel DNA-control element. Mol Cell Biol 16: 1479–1489

Varmus HE (1984) The molecular genetics of cellular oncogenes. Annu Rev Genet 18: 553–612

Wegner M, Cao Z, Rosenfeld M (1991) Calcium-regulated phosphorylation within the leucine zipper of C/EBPβ. Science 256: 370–373

Williams SC, Cantwell CA, Johnson PF (1991) A family of C/EBP-related proteins capable of forming covalently linked leucine zipper dimers in vitro. Genes Dev 5: 1553–1567

Zhan Q, Liebermann DA, Alamo I, Hollander MC, Ron D, Kohn KW, Hoffman B, Fornace AJ (1994) The gadd and MyD genes define a novel set of mammalian genes encoding acidic proteins that cooperatively suppress cell growth. Mol Cell Biol 14: 2361–2371

Zinszner H, Albalat R, Ron D (1994) A novel effector domain from the RNA-binding proteins TLS or EWS is required for oncogenic transformation by CHOP. Genes Dev 8: 2513–2526

Zucman J, Delattre O, Desmaze C, Epstein AL, Stenman G, Spelman F, Fletchers CDM, Aurias A, Thomas G (1993a) EWS and ATF-1 gene fusion induced by t(12;22) translocation in malignant melanoma of soft parts. Nat Gen 4: 341–345

Zucman J, Melot T, Desmaze C, Ghysdeal J, Plougastel B, Peter M, Zucker JM, Triche TT, Sheer D, Turc-Carel C, Ambros P, Combaret V, Lenoir G, Aurias A, Thomas G, Delattre O (1993b) Combinatorial generation of variable fusion proteins in the Ewing family of tumors. EMBO J 12: 4481–4487

Biology of EWS/FLI and Related Fusion Genes in Ewing's Sarcoma and Primitive Neuroectodermal Tumor

W.A. May[1] and C.T. Denny[1,2]

1 Ewing's Sarcoma and Primitive Neuroectodermal Tumor Cells Contain Characteristic Cytogenetic Abnormalities Involving Chromosome 22

Ewing's sarcoma and peripheral primitive neuroectodermal tumor (ES/PNET) are two related tumors of bone and soft tissue. Both malignancies belong to the enigmatic diagnostic category of small round cell tumors of childhood, which as a group occurs at a rate of 29 per million children (Triche et al. 1987). The precise cell of origin of ES/PNET is unknown. Circumstantial evidence suggests that these tumors may arise from neuroectodermal precursors (for review see Cavazzana et al. 1992). For example, ES/PNET cells express high levels of choline acetyl transferase, an enzyme important in the biosynthesis of cholinergic neurotransmitters. In addition, some ES/PNET cell lines can be induced in culture to form neurite extensions and express neurofilament proteins.

On a practical level, the diagnosis of ES/PNET has rested more on the lack of identifiable phenotypic features rather than on clear specific markers. Surprisingly, this seemingly arbitrary process of elimination has resulted in a group of tumors which exhibit a common karyotypic abnormality. Approximately 85% of

[1]Department of Pediatrics, Gwynne Hazen Cherry Memorial Laboratories, Division of Hematology/Oncology, and Jonsson Comprehensive Cancer Center, School of Medicine, University of California, Los Angeles, CA 90024, USA
[2]Department of Pediatrics, A2-312 MDCC, UCLA Medical Center, Los Angeles, CA 90095, USA

ES/PNET tumors contain a cytogenetically detectable translocation between chromosomes 11 and 22: t(11;22)(q24;q12) (WHANG-PENG et al. 1986; TURC-CAREL et al. 1988). An additional 5%–10% of tumors contain an alternative rearrangement juxtaposing chromosomes 21 and 22: t(21;22)(q21;q12) (ZUCMAN et al. 1993; SORENSEN et al. 1994). Finally there have been sporadic reports of other ES/PNET cytogenetic translocations involving the same region of chromosome 22 fused to other partners such as chromosome 7: t(7;22)(q22;q12) (JEON et al. 1995). Molecular cloning and characterization of these translocation breakpoints (DELATTRE et al. 1992; ZUCMAN et al. 1992) have provided new tools for unambiguous diagnosis of ES/PNET (DOWNING et al. 1993; SORENSEN et al. 1993; TAYLOR et al. 1993; TORETSKY et al. 1995) and has opened new avenues for research into the biology of these tumors.

2 Ewing's Sarcoma/Primitive Neuroectodermal Tumor Translocations Fuse the EWS Gene on Chromosome 22 to Members of the ETS Family of Transcription Factors

The breakpoints and neighboring genes of the ES/PNET t(11;22) were isolated by positional cloning strategies (DELATTRE et al. 1992). Molecular analysis of this rearrangement revealed that the t(11;22) formed a fusion between the NH_2-terminal region of a gene named *EWS* and the COOH-terminal of *FLI-1*, a member of the ETS family of transcription factors (Fig.1). In similar fashion, the variant t(21;22) and t(7;22) rearrangements were subsequently shown to fuse approximately the same regions of EWS to the carboxyl domains of the ETS genes *ERG* (ZUCMANN et al. 1993; DUNN et al. 1994; GIOVANNINI et al. 1994; SORENSEN et al. 1994) and *ETV-1* (JEON et al. 1995), respectively.

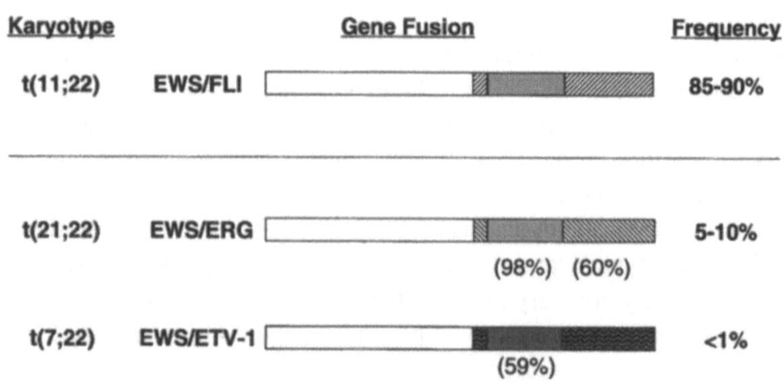

Fig. 1. Chimeric fusion genes found in ES/PNET. Characteristic chromosomal translocations fuse the NH_2-terminal portion of the EWS gene (*clear*) to the COOH terminals of one of several ETS transcription factors: FLI-1 (*hatch*); ERG (*reverse hatch*); ETV-1 (*wave*). ETS common domains that mediate sequence-specific DNA-binding are shown in gray. The percent amino acid identity with FLI-1 for ERG and ETV-1 is shown in *parentheses*

EWS was discovered by its association with ES/PNET tumors. While its normal cellular function remains to be established, accumulating evidence suggests that EWS may be involved in RNA processing. Sequence analyses in the COOH-terminal portion of EWS revealed regions of similarity to known RNA-binding proteins (DELATTRE et al. 1992). This same EWS domain is also structurally related to the TLS/FUS gene, which is fused to other transcription factors as a result of specific chromosomal translocations found in particular human tumors (see chapter by RON, this volume). Moreover, it has been shown that the COOH terminal half of EWS can mediate RNA binding in vitro (OHNO et al. 1994).

It is approximately the NH_2-terminal 285 residues of EWS which are fused to FLI-1 by the t(11;22). This region consists of a series of degenerate repeats rich in glutamine, arginine, and proline. While this region has modest homology to the COOH- terminal domain of RNA polymerase II, the structural significance of this sequence homology is uncertain (DELATTRE et al. 1992). However, its glutamine-rich amino acid composition suggests that this domain could potentially function as a transcriptional activation domain.

FLI-1, the other partner in the t(11;22), was originally discovered by virtue of its activation by Friend leukemia virus integration in murine erythroleukemia (BEN-DAVID et al. 1991). As with all ETS transcription factors, it contains a common ETS domain which mediates sequence-specific DNA binding. In the case of FLI-1, this region is located at the COOH-terminal portion of the molecule (KLEMSZ et al. 1993; ZHANG et al. 1993). Both structural analyses and model reporter gene assays suggest that the NH_2-terminal portion of FLI-1 functions as a transcriptional activation domain (WATSON et al. 1992; ZHANG et al. 1993; KLEMSZ et al. 1993). Deletional mutagenesis studies further suggest that there may be an additional transcriptional activation domain located in the extreme COOH-terminal of FLI-1 (RAO et al. 1993). While some of the specificity of action of ETS proteins can be explained by DNA binding specificity, a variety of ETS proteins can be demonstrated to interact with a given binding site (for review see JANKNECHT and NORDHEIM 1993). Protein-protein interactions with other transcription complex members are thought to enhance specificity of action, as observed with ETS-1 and Sp1 (GEGONNE et al. 1993).

As a result of the t(11;22) the putative transcriptional activation domain of FLI-1 is replaced by the glutamine-rich NH_2-terminal sequences of EWS. As with many translocation associated chimeric molecules, there is variability seen in different fusion products found in different tumors (ZUCMAN et al. 1993). In general, this reflects differences in chromosomal breakpoints. In the t(11;22), breakpoints on chromosome 22 occur within a 7 kb region bounded by exons 7 and 11 of EWS, while those on chromosome 11 occur over a 50 kb range extending from intron 3 to intron 7 of FLI-1 (ZUCMAN et al. 1992). To compound the complexity, some ES/PNET tumor cell lines express more than one chimeric fusion, probably as a result of differential RNA splicing (MAY et al. 1993a). While these structural differences are evident, their biologic importance is less clear. Thus far, precise fusion points have not been shown to have any correlation with tumor phenotype or clinical parameters (DELATTRE et al. 1994; ZOUBEK et al. 1994).

By contrast, the invariant inclusion of the first seven exons of EWS coupled to the DNA binding domain of FLI-1 strongly suggests that EWS/FLI may function as aberrant transcription factors in the development of ES/PNET. These same essential elements of the EWS/FLI chimera are recapitulated in the variant ES/PNET fusions. The t(21;22) forms a chimera fusing the same NH_2- terminal EWS region to ERG, an ETS transcription factor highly homologous to FLI-1. The resultant EWS/ERG fusion is structurally very similar to EWS/FLI and, as seen before, the ETS domain of ERG is always conserved (Zucman et al. 1993; Sorensen et al. 1994). Recently a more distantly related ETS transcription factor, ETV-1, has been shown to fuse to EWS in rare cases exhibiting a variant translocation, t(7;22) (p22;q12) (Jeon et al. 1995). The striking similarity of the products of these variant ES/PNET chimera suggests that may function similarly.

3 *EWS/FLI* and *EWS/ERG* Are Dominant Acting Oncogenes

Some 95% of ES/PNET express EWS chimera of one type or another (Delattre et al. 1994), a fact which suggests a key role for these chimeric proteins in the pathogenesis of these tumors. This critical role is supported by studies using model transformation systems. NIH 3T3 cells expressing either EWS/FLI or EWS/ERG efficiently form colonies in soft agar (May et al. 1993a). Furthermore, subcutaneous injection of EWS/FLI expressing NIH 3T3 cells into nude mice results in gross tumor formation within 2 weeks. In contrast, normal FLI-1 does not transform NIH 3T3 cells (May et al. 1993b). These data suggest that EWS/FLI is a molecule that is functionally distinct from FLI-1.

Transformation by EWS/FLI requires both EWS and FLI-1 domains. Deletion of either EWS or the ETS domain of FLI-1 completely abrogates transforming activity (May et al. 1993a). The requirement for both components of the EWS/FLI chimera suggests that EWS/FLI has a dominant mode of action and does not act primarily by interfering with either normal cellular EWS or FLI-1.

EWS/FLI does not transform all cells. Neither Rat1 cells nor particular strains of NIH 3T3 cells are transformed by EWS/FLI. This indicates that EWS/FLI must be expressed in a permissive cellular background in order for transformation to occur and suggests that EWS/FLI interacts with other cellular factors. At least two scenarios are possible: (1) transformation competent cells express necessary genes that are required for EWS/FLI transformation; (2) transformation resistant cells express proteins that inhibit the effect of EWS/FLI.

4 *EWS/FLI* Has Biochemical Characteristics of a Transcription Factor

Several observations suggest that EWS/FLI functions as an aberrant transcription factor: (1) FLI-1, one partner in the chimeric fusion, is a known transcription factor; (2) the FLI-1 DNA-binding domain is invariably present in all EWS/FLI fusions isolated from ES/PNET cells; (3) the FLI-1 DNA-binding domain is required for transformation by EWS/FLI. Strengthening this hypothesis, EWS/FLI displays particular biochemical characteristics that are common to transcription factors. First, the EWS/FLI protein localizes to the nucleus in ES/PNET cells (MAY et al. 1993b; BAILLY et al. 1994). Second, EWS/FLI can bind DNA in a site-specific manner (MAY et al. 1993b; OHNO et al. 1993; BAILLY et al. 1994). In fact, EWS/FLI and normal FLI-1 manifest very similar DNA-binding characteristics (MAO et al. 1994). Third, it also appears that the portion of EWS that displaces the NH_2-terminals of FLI-1 and ERG encodes a potent transcriptional activation domain. When coupled to a yeast GAL4 DNA-binding domain, EWS is able to activate model reporter genes much more efficiently than the corresponding transcriptional activation domain of FLI-1 (MAY et al. 1993b; BAILLY et al. 1994). In similar fashion, intact EWS/FLI and EWS/ERG fusions have been shown to transcriptionally activate reporter constructs (OHNO et al. 1993, 1994). Finally, known, structurally distinct transcriptional activation domains can functionally replace the EWS domain in the EWS/FLI fusion and result in chimeric FLI-1 fusions that can transform NIH 3T3 cells (LESSNICK et al. 1995).

How then does EWS/FLI transform cells that normal FLI-1 cannot transform? A particularly attractive hypothesis is that EWS/FLI is able to transcriptionally modulate a set of target genes in a manner different from FLI-1. Even though EWS/FLI and FLI-1 may bind the same DNA sites, their abilities to transcriptionally activate specific target genes may be very different. ETS proteins have been shown to form heteromeric complexes with other transcription factors during the activation of target genes. These intermolecular protein-protein interactions are critical to the specificity of certain ETS proteins (for review see WASYLYK et al. 1993). For example, the ETS protein SAP-1 requires an additional factor (SRF) to productively activate the c-*FOS* gene (DALTON and TREISMAN 1992). FLI-1, like other ETS factors, may normally function in conjunction with other requisite cofactors. If this is true, replacement of the normal transcriptional activation domain with the potentially more potent EWS domain could have significant functional consequences. EWS/FLI may not be as constrained in its target selection and could potentially up-regulate genes without cofactors that are essential for FLI-1.

5 Future Questions

The hypothesis that EWS/FLI and related chimeras transform cells by acting as aberrant transcription factors to inappropriately modulate gene expression prompts the basic question: What are the target genes? Using a candidate gene approach, one study has suggested that the c-*MYC* gene may be up-regulated as a result of EWS/FLI expression (Bailly et al. 1994). More recently, the technique of representational difference analysis has been employed to identify genes which are differentially expressed in NIH 3T3 cells containing EWS/FLI or FLI-1 (Braun et al. 1995). A cadre of genes that is transcriptionally activated by EWS/FLI but not FLI-1 has been identified providing further evidence that EWS/FLI does specifically alter gene expression. These and other EWS/FLI target genes coupled with model transformation systems will allow the task of explicitly defining a molecular transformation pathway in ES/PNET to be addressed.

The study of the *EWS/FLI* fusion gene is likely to have direct clinical impact as well. The diagnosis of ES/PNET needs no longer to be one of exclusion but can now be based in the presence of a true tumor-linked marker. Already the unexpected presence of EWS/FLI-1 in sarcomas demonstrating both myogenic and neural differentiation serves as an example of how tumors, thought to be previously unrelated, may be biologically linked (Sorensen et al. 1995). Minimal disease detection schemes using RT-PCR assays specific for EWS/FLI can now be constructed and hopefully will provide useful adjuncts to the treatment of patients with this aggressive malignancy. Finally, defining the function and the transformation pathways of the EWS/FLI fusion gene may very well lead to the development of specific antitumor strategies.

References

Bailly RA, Bosselut R, Zucman J, Cormier F, Delattre O, Roussel M, Thomas G and Ghysdael J (1994) DNA-binding and transcriptional activation properties of the fusion protein resulting from the t(11;22) translocation in Ewing sarcoma. Mol Cell Biol 14: 3230–3241

Ben-David YEB, Giddens EB, Letwin K and Bernstein A (1991) Erythroleukemia induction by Friend murine leukemia virus: insertional activation of a new member of the *ETS* gene family *FLi-1*, closely linked to c-*ETS*-1. Genes Dev 5: 908–918

Braun BS, Freiden R, Lessnick SL, May WA and Denny CT (1996) Identification of target genes to the Ewing's sarcoma EWS/FLI fusion protein by representational difference analysis. Mol Cell Biol (in press)

Cavazzana AO, Ninfo V, Roberts J and Triche TJ (1992) Peripheral neuroepithelioma: a light microscopic, immunocytochemical, and ultrastructural study. Mod Pathol 5: 71–78

Dalton S and Treisman R (1992) Characterization of SAP-1, a protein recruited serum response factor to the *c-fos* serum response element. Cell 68: 597–612

Delattre O, Zucman J, Melot T, Garau XS, Zucker JM, Lenoir GM, Ambros PF, Sheer D, Turc-Carel C, Triche TJ et al. (1994) The Ewing family of tumors – a subgroup of small-round-cell by specific chimeric transcripts. N Engl J Med 331: 294-299

Delattre O, Zucman J, Plougastel B, Desmaze C, Melot T, Peter M, Kovar H, Joubert I, de Jong P, Rouleau G, Aurias A and Thomas G (1992) Gene fusion with an ETS DNA-binding domain caused by translocation in human tumours. Nature 359: 162–165

Downing JR, Head DR, Parham DM, Douglass EC, Hulshof MG, Link MP, Motroni TA, Grier HE, Curcio-Brint AM and Shapiro DN (1993) Detection of the (11;22)(q24;q12) translocation of Ewing's peripheral neuroectodermal tumor by reverse transcription chain reaction. Am J Pathol 143: 1294–1300

Dunn T, Praissman L, Hagag N and Viola MV (1994) ERG gene is translocated in an Ewing's sarcoma cell line. Cancer Genet Cytogenet 76: 19–22

Gegonne A, Bosselut R, Bailly RA and Ghysdael J (1993) Synergistic activation of the HTLV1 LTR ETS-responsive region by transcription factors Ets1 and Sp1. EMBO J 12: 1169–1178

Giovannini M, Biegel JA, Serra M, Wang JY, Wei YH, Nycum L, Emanuel BS and Evans GA (1994) EWS-erg and EWS-Fli1 fusion transcripts in Ewing's sarcoma neuroectodermal tumors with variant translocations. J Clin Invest 94: 489–496

Janknecht R and Nordheim A (1993) Gene regulation by ETS proteins. Biochim Biophys Acta 1155: 346–356

Jeon IS, Davis JN, Braun BS, Sublett JE, Roussel MF, Denny CT and Shapiro DN (1995) A variant Ewing's sarcoma translocation (7;22) fuses the EWS ETS gene ETV1. Oncogene 10: 1229–1234

Klemsz MJ, Maki RA, Papayannopoulou T, Moore J and Hromas R (1993) Characterization of the ETS oncogene family member, fli-1. J Biol Chem 268: 5769–5773

Lessnick SL, Braun BS, Denny CT and May WA (1995) Multiple domains mediate transformation by the Ewing's sarcoma EWS/FLI-1 fusion gene. Oncogene 10: 423–431

Mao X, Miesfeldt S, Yang H, Leiden JM and Thompson CB (1994) The FLI-1 and chimeric EWS-FLI-1 oncoproteins display similar specificities. J Biol Chem 269: 18216–18222

May WA, Gishizky ML, Lessnick SL, Lunsford LB, Lewis BC, Delattre O, Zucman J, Thomas G and Denny CT (1993a) Ewing sarcoma 11;22 translocation produces a chimeric factor that requires the DNA-binding domain encoded by FLI1 for transformation. Proc Natl Acad Sci USA 90: 5752–5756

May WA, Lessnick SL, Braun BS, Klemsz M, Lewis BC, Lunsford LB, Hromas R and Denny CT (1993b) The Ewing's sarcoma EWS/FLI-1 fusion gene encodes a more transcriptional activator and is a more powerful transforming FLI-1. Mol Cell Biol 13: 7393–7398

Ohno T, Ouchida M, Lee L, Gatalica Z, Rao VN and Reddy ES (1994) The EWS gene, involved in Ewing family of tumors, malignant soft parts and desmoplastic small round cell tumors, codes for binding protein with novel regulatory domains. Oncogene 9: 3087–3097

Ohno T, Rao VN and Reddy ES (1993) EWS/Fli-1 chimeric protein is a transcriptional activator. Cancer Res 53: 5859–5863

Rao VN, Ohno T, Prasad DD, Bhattacharya G and Reddy ES (1993) Analysis of the DNA-binding and transcriptional activation human Fli-1 protein. Oncogene 8: 2167–2173

Sorensen PH, Shimada H, Liu XF, Lim JF, Thomas G and Triche TJ (1995) Biphenotypic sarcomas with myogenic and neural the Ewing's sarcoma EWS/FLI1 fusion gene. Cancer Res 55: 1385–1392

Sorensen PH, Lessnick SL, Lopez-Terrada D, Liu XF, Triche TJ and Denny CT (1994) A second Ewing's sarcoma translocation, t(21;22), fuses the another ETS-family transcription factor, ERG. Nat Genet 6: 146–151

Sorensen PH, Liu XF, Delattre O, Rowland JM, Biggs CA, Thomas G and Triche TJ (1993) Reverse transcriptase PCR amplification of EWS/FLI-1 fusion as a diagnostic test for peripheral primitive neuroectodermal childhood. Diagn Mol Pathol 2: 147–157

Taylor C, Patel K, Jones T, Kiely F, De Stavola BL and Sheer D (1993) Diagnosis of Ewing's sarcoma and peripheral neuroectodermal on the datection of t(11;22) using fluorescence in situ. Br J Cancer 67: 128–133

Toretsky JA, Neckers L and Wexler LH (1995) Detection of (11;22) (q24;q12) translocation-bearing cells in blood progenitor cells of patients with Ewing's sarcoma family. J Natl Cancer Inst 87: 385–386

Triche TJ, Askin FB, Kissane JM (1987) Neuroblastoma, Ewing's sarcoma, and the differential diagnosis of small-, round-, blue-cell tumors. In: Feingold M, Benningtion JC (eds) Major problems in pathology, vol 18. Saunders, Philadelphia, pp 145–195

Turc-Carel C, Aurias A, Mugneret F, Lizard S, Sidaner I, Volk C, Thiery JP, Olschwang S, Philip I, Berger MP et al. (1988) Chromosomes in Ewing's sarcoma. I. An evaluation of 85 cases of remarkable consistency of t(11;22) (q24;q12). Cancer Genet Cytogenet 32: 229–238

Wasylyk B, Hahn SL and Giovane A (1993) The ETS family of transcription factors. Eur J Biochem 211: 7–18

Watson DK, Smyth FE, Thompson DM, Cheng JQ, Testa JR, Papas TS, Seth A (1992) The *ERGB/FLI-1* gene: isolation and characterization of a new member of the family of human *ETS* transcription factors. Cell Growth Differ 3: 705–713

Whang-Peng J, Triche TJ, Knutsen T, Miser J, Kao-Shan S, Tsai S and Israel MA (1986) Cytogenetic characterization of selected small round cell tumors of childhood. Cancer Gene Cytogene 21: 185–208

Zhang L, Lemarchandel V, Romeo PH, Ben-David Y, Greer P and Bernstein A (1993) The Fli-1 proto-oncogene, involved in erythroleukemia and sarcoma, encodes a transcriptional activator with DNA-binding specificities distinct from other ETS family members. Oncogene 8: 1621–1630

Zoubek A, Pfleiderer C, Salzer-Kuntschik M, Amann G, Windhager R, Fink FM, Koscielniak E, Delattre O, Strehl S, Ambros PF et al. (1994). Variability of EWS chimaeric transcripts in Ewing tumours: a of clinical and molecular data. Br J Cancer 70: 908–913

Zucman J, Delattre O, Desmaze C, Plougastel B, Joubert I, Melot T, Peter M, De Jong P, Rouleau G, Aurias A et al. (1992) Cloning and characterization of the Ewing's sarcoma and neuroepithelioma t(11;22) translocation breakpoints. Genes Chromosom Cancer 5: 271–277

Zucman J, Melot T, Desmaze C, Ghysdael J, Plougastel B, Peter M, Zucker JM, Triche TJ, Sheer D, Turc-Carel C et al. (1993). Combinatorial generation of variable fusion proteins in the Ewing family of tumours. EMBO J 12: 4481–4487

Chromosome Translocation-Mediated Conversion of a Tumor Suppressor Gene into a Dominant Oncogene: Fusion of EWS1 to WT1 in Desmoplastic Small Round Cell Tumors

F.J. RAUSCHER, III

1 Overview

The preceding articles in this volume have provided a unique overview of genetic and biochemical mechanisms which underlie oncogenic conversion of transcription factor function. As is evident, many of the models have derived from the study of pediatric and adult leukemias. These studies have provided a paradigm for the emerging analyses of chromosomal translocations involving transcription factors in solid tumors. This is exemplified by the manuscript of Barr who examines the PAX3-FKHR fusion transcription factor whose mechanism likely involves increased activation of normal PAX3 target genes. A similar but probably not identical mechanistic theme is echoed by the EWS and TLS fusions described by RON and MAY. Like PAX3-FKHR, the EWS/TLS fusions confer a novel activation domain to an otherwise unaltered DNA binding domain, thereby creating a dominant oncogene. Most remarkable is the diversity of DNA-binding domain types which are involved in EWS/TLS fusions and, concomitantly, the diversity of the disease processes initiated. This article will expand on this theme by describing a new member of the EWS/TLS family of oncogenes, namely EWS-WT1, which occurs in the solid tumor desmoplastic small round cell sarcoma (DSRCT). The EWS–WT1 fusion is unique in that its DNA binding domain is derived from the Wilms' tumor-1 (WT1) tumor suppressor protein. This is one of the first examples of chromosomal-translocation-mediated fusion of a proto-oncogene (EWS) and a

The Wistar Institute, 3601 Spruce Street, Philadelphia, PA 19104, USA. Phone: (215) 898-0995; Fax: (215) 898-3929; e-mail: rauscher@wista.wistar.upenn.edu tumor suppressor gene (WT1) which creates a dominant oncogene. In order to understand the context of these findings, we will first provide brief reviews of the biology and genetics of WT1, EWS, and DSRCT.

2 WT1 and Wilms' Tumorigenesis

Wilms' tumor is pediatric nephroblastoma which occurs in children with a frequency of approximately 1/10 000 live births. The association of Wilms' tumor with cytogenetic deletions of chromosome 11 and developmental malformations in the Wilms' tumor, aniridia, genitourinary abnormalities and mental retardation (WAGR) syndrome (reviewed in HABER et al. 1992) formed the basis for the identification of the WT1 Wilms' tumor gene on chromosome 11p13. WT1 is a tumor suppressor gene and, consistent with this, *wt1* is mutated or deleted in a subset of Wilms' tumors (reviewed in COPPES et al. 1993). The properties of the WT1 gene and protein are summarized in Table 1. WT1 is expressed at the highest levels primarily during development of the kidney and genitourinary systems (PRITCHARD-JONES et al. 1990), but is also expressed in the developing gonad, spleen, and mesothelium (reviewed in HABER and HOUSMAN 1992). Targeted disruption of the murine *wt1* gene resulted in embryonic lethality that was associated with a failure of kidney and gonadal development (KREIDBERG et al. 1993), further substantiating the critical role played by WT1 during nephrogenesis and overall urogenital system development. The biological importance of the gene is also demonstrated by the preponderance of constitutional *wt1* mutations that lead to severe urogenital abnormalities and Wilms' tumor in Denys-Drash syndrome (PELLETIER et al. 1991).

The *wt1* mRNA is alternatively spliced, giving rise to four transcripts and reflecting the presence or absence of two exons (HABER et al. 1990, 1991). One alternatively spliced exon, exon 5, encodes a 17 amino acid segment in the amino terminus of the protein, whereas the second results in the insertion of three amino acids, lysine, threonine and serine (abbreviated KTS), between the third and fourth zinc fingers (HABER et al. 1990, 1991). The relative abundance of each of the splice variants is consistent in both the developing kidney as well as in Wilms' tumors that have been examined, with the WT1(+KTS) isoform being about four times as abundant as the WT1 (−KTS) form (HABER et al. 1991). WT1(+KTS) does not bind to the early growth response (EGR) core consensus (RAUSCHER et al. 1990). However, a DNA-binding site has been identified for WT1(+KTS) which contains the EGR-1 core consensus, but requires additional flanking sequences for high-affinity binding (DRUMMOND et al. 1994). The WT1(+KTS) protein has also been shown to be physically associated with the splicing apparatus (LARSSON et al. 1995), suggesting that WT1 may play a post-transcriptional role in gene regulation.

The *wt1* gene encodes a DNA-binding protein, WT1, which contains a proline- and glutamine-rich amino terminus and four zinc fingers of the Cys_2-His_2 class at the carboxy terminus (CALL et al. 1990). The WT1 protein recognizes the core

Table 1. Summary of genetic biological and biochemical properties for the WT1, Wilms' tumor suppressor gene and protein

Properties	Description
Location	Chromosome 11p13
Size	~50 kilobase genomic locus
Structure	10 exons, 2 alternative splices
mRNA	~3.5 kb
Expression patterns	Embryonic - kidney (condensing metanephric blastema and podocytes) - mesothelial lining (all organs) - gonadal ridge mesothelium - spleen - brain (area postrema) - spinal cord (ventral horn motor neurons) Adult - kidney (glomerular epithelium) - ovary (granulosa cells) - testis (Sertoli cells) - uterus (decidual cells)
Protein product	52–54 kDa, nuclear protein
Structural motifs	4 Cys_2-His_2 zinc fingers, glutamine-proline-glycine-rich transcriptional regulation domain
Interacting proteins	p53
DNA binding site	EGR consensus sequence: 5′-GGAGCGGGGGCG-3′
Target genes	*IGF-II*, *IGF-II*-receptor, *Egr-1*, *Pax-2*, *PDGF-A*, *CSF-1*, *TGF-beta1*
Diseases associated with	Wilms' tumor, Denys-Drash syndrome, mesothelioma, desmoplastic small round cell tumor

The studies summarized here are referenced in the text

sequence 5′-GCGGGGGCG-3′, the DNA consensus binding site for the members of the immediate early EGR family of proteins (Rauscher et al. 1990). When the EGR proteins are bound to this site, they often activate the transcription of downstream target genes (reviewed in Sukhatme 1990), whereas when WT1 is bound to this site, it often functions as a transcriptional repressor (Madden et al. 1991). Consistent with WT1 having a role as a tumor suppressor, it has recently been shown to regulate the transcriptional activity of a number of genes whose products are involved in the promotion of growth. These include the genes for the insulin-like growth factor II (*IGFII*) (Drummond et al. 1992), insulin-like growth factor 1 receptor (*IGF-I* rec) (Werner et al. 1994), platelet-derived growth factor α-chain (*PDGF-α*) (Wang et al. 1992; Gashler et al. 1992), transforming growth factor-β (*TGF-β*) (Dey et al. 1994), the retinoic acid receptor-α (*RAR-α*) (Goodyer et al. 1995) the paired-box gene *PAX-2* (Ryan et al. 1995), and *syndecan-1* (Cook et al. 1996). Several of these genes show an expression pattern similar to that of *wt1* in the developing kidney and are overexpressed in Wilms' tumors, suggesting they are physiologically relevant downstream target genes for WT1.

Wilms' tumors are often associated with the persistence of undifferentiated metanephric mesenchyme, or nephrogenic rests (Beckwith et al. 1990). Histological examination of Wilms' tumors has revealed that cells within the tumor show varying degrees of differentiation resembling various stages of nephrogenesis which have arrested prematurely (Beckwith et al. 1990). The human kidney develops through a reciprocal induction between the metanephric mesenchyme and the invading epithelial ureteric bud (reviewed in Saxen 1987). Within the developing kidney, *wt1* expression increases dramatically when the mesenchymal cells condense around the ureteric buds (Pritchard-Jones et al. 1990; Pelletier et al. 1991). The overall picture of WT1 expression during nephrogenesis suggests it plays a crucial role in the transdifferentiation event of mesenchyme to epithelium. More importantly, WT1 appears to have a broad role in mesothelial cell function as defined by the following observations: (1) WT1 is expressed at high levels in the peritoneal mesothelium in both embryonic development and the adult; (2) mesotheliomas of the lung have been shown to carry mutations in WT1 which alter its function as a transcriptional repressor; and (3) WT1 knockout mice display evidence of abnormal peritoneal and pleural mesothelia which contribute to heart and lung abnormalities possibly leading to the embryonic lethality observed. As described below, these observations combined with the fact that (1) DSRCT appear to arise from transformation of the peritoneal mesothelium and (2) DSRCT contain cytogenetic abnormalities involving the 11p13 WT1 chromosomal locus allowed us to infer an involvement of WT1 in this tumorigenic process.

3 EWS1 and Ewings' Sarcoma

Ewings' sarcoma occurs in bone and soft tissue and is highly malignant in both children and adults. The tumor is often characterized histologically by sheets of uniform small round cells. These small round cell tumors (SRCT) include Ewings' sarcoma, embryonal/alveolar rhabdomyosarcomas, Askins' tumor and peripheral neuroectodermal tumor (reviewed in Delattre et al. 1994) and are characterized by recurrent chromosomal translocations involving the *EWS1* gene (which encodes a putative RNA binding transcription factor) on chromosome 22q12 and a number of other loci depending on tumor type (Fig. 1). The wild-type *EWS1* gene encodes a novel protein with transcription-factor-like characteristics (Fig. 1): an RNA binding domain homologous to hnRNP proteins, amino acid segments rich in proline, arginine or glycine, and an NH_2-terminal domain (NTD) comprised of ~31 repeats of the hexapeptide – SYSQQS – reminiscent of the COOH-terminal domain of RNA polymerase II (Delattre et al. 1992). The NTD displays ~50% amino-acid sequence homology to the *TLS/FUS* (see Ron in this volume) gene which is translocated to *CHOP*, a bZIP domain transcription factor myxoid liposarcoma (Crozat et al. 1993; Rabbits et al. 1993). When involved in a translocation, the NTD of *EWS1* is fuse in-frame to a variety of DNA binding domains: (1) the ETS

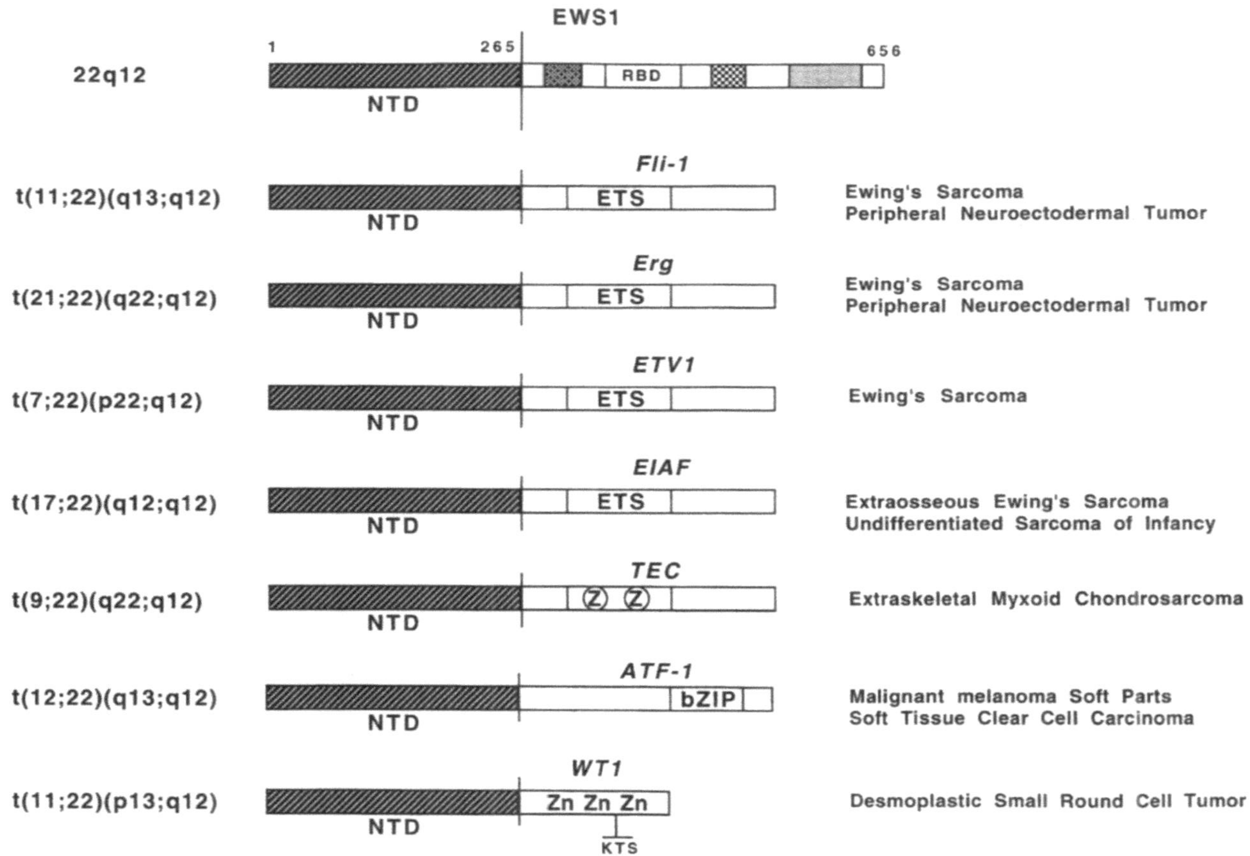

Fig. 1. Summary of the chromosomal translocations involving the EWS1, Ewings' sarcoma gene. The recurrent cytogenetic abnormalities in the indicated malignancies are shown. In each case, a fusion transcript and/or protein has been identified between the NH_2-terminal domain (NTD) of EWS1 and the indicated DNA binding domain. RBD, RNA binding domain

class DNA binding domains of *Fli-1* [t(11;22)] (Delattre et al. 1992), *Erg-1* [t(21;22)] (Sorenson et al. 1994), or ETV1 [t(7;22)] (Jeon et al. 1995) in Ewings' sarcoma and peripheral neuroectodermal tumors; (2) the E1AF ETS domain protein [(17;22)] in extraosseous Ewings' sarcoma and undifferentiated sarcoma of kidney (Kaneko et al. 1996, Urano et al. 1996); (3) the bZIP DNA binding domain of *ATF-1* [t(12;22)] (Zucman et al. 1993) in soft tissue clear cell sarcoma and malignant melanoma of soft parts; and (4) the nuclear hormone receptor zinc finger type DNA binding domain from TEC [t(9;22)]; an orphan receptor (Labelle et al. 1995) in extraskeletal myxoid chondrosarcoma. It is remarkable that each translocation involves fusion of a highly potent transcriptional activation NTD from EWS to a novel DNA binding domain and creates a potent tumor-type-specific chimeric transcription factor with oncogenic potential. As described in May and Denny (this volume), the EWS-DBD fusions are dominant oncogenes, the transforming activity of which are dependent upon the EWS activation domain. With this paradigm in mind, we have molecularly characterized a newly recognized clinicopathologic subtype of SRCT, namely DSRCT, which also contains a recurrent chromosomal translocations involving the 22q12 *EWS1* locus. As described below, we and others have shown that the *EWS1* gene is fused to the zinc finger region of the *WT1* Wilms' tumor suppressor gene as a result of the t(11;22) (p13;q12) in DSRCT (Fig. 1).

4 Desmoplastic Small Round Cell Tumor and the EWS–WT1 Fusion

DSRCT (also known as intra-abdominal DSRCT) (Ordonez et al. 1989, 1993; Gerald et al. 1991) is a highly aggressive, rare tumor which occurs most frequently in adolescent males and is located almost exclusively in the abdomen (Table 2). At presentation, patients often have tumor involvement of many abdominal organs in the abdomen and also in the mesothelial lining of the gut. This property of the tumor has hampered both identification of the primary site of tumor development and the target cell for oncogenic transformation. A "nested" pattern of tumor cell growth containing islands of densely packed small round cells amongst the characteristic (and diagnostic) desmoplastic stroma is notable histologically. Re-

Table 2. Summary of clinical and pathogenetic properties of desmoplastic small round tumors

1. *Most frequent in adolescent males*
2. *Almost exclusive intra-abdominal location*
3. *Nested pattern of cell growth*
4. *Intense desmoplastic reaction*
5. *Express epithelial, mesenchymal and neural cell markers*
6. *recurrent t(11;22)(p13;q12)*

markably, immunohistochemical analysis demonstrates that DSRCT often co-express mesenchymal, epithelial and neural markers. Together, these findings, have led pathologists to consider other designations for this tumor based on this primitive cellular phenotype, most notably peritoneal blastoma or extra-renal Wilms' tumor. This last designation stems from similarities amongst DSRCT and "classic" triphasic Wilms' tumors which also contain epithelial, mesenchymal, and stromal cell elements (BECKWITH et al. 1990). A key finding in the link between DSRCT, EWS1, and WT1 were tumor cytogenetics; a number of investigators independently identified a translocation t(11;22)(p13;q12) in DSRCT (SAWYER et al. 1992; SHEN et al. 1992; BIEGEL et al. 1993; RODRIGUEZ et al. 1993) possibly involving the 22q12 *EWS1* gene and the *WT1* gene at 11p13. Thus, based on (1) cytogenetics, (2) similar histopathologic profiles of triphasic Wilms' and DSRCT, (3) the knowledge that the mesothelia is a site of *WT1* expression, and (4) that *EWS1* is always fused to a DNA binding domain when involved in a chromosomal translocations (Fig. 1), we characterized *WT1* and *EWS1* genes in DSRCT.

Since there are no cell lines for DSRCT, we utilized frozen tumor specimens or, in most cases, fixed specimens in paraffin blocks. Our experimental strategy was to hypothesize that a fusion mRNA occurred between the NTD of EWS1 and the zinc finger region of WT1. We developed a PCR assay for this fusion transcript which detected a ~100 base pair PCR product from DSRCT tumor specimens using primers specific for both EWS1 and WT1. Cloning and sequencing of this product demonstrated fusion of EWS1 exon 7 to WT1 exon 8 (which encodes the zinc finger 2 of the WT1 DNA binding domain) (RAUSCHER et al. 1994). To date, more than 20 cases which were assigned the clinical diagnosis of DSRCT have been analyzed using this PCR assay; 17 of these have been positive for the EWS–WT1 fusion transcript. Thus, the presence of an EWS–WT1 fusion is a common feature of independent DSRCT. Recently, this procedure has been developed into a specific diagnostic assay (DE ALAVA et al. 1995), which allowed highly specific and differential diagnosis of DSRCT from other small round cell sarcomas. Additional supportive data for alterations of EWS and WT1 loci include rearrangements at the genomic level via Southern blotting (LADANYI and GERALD 1994; RAUSCHER et al. 1994; and unpublished data). Furthermore, we have detected the EWS–WT1 fusion polypeptide in tumor specimens using a combination of immunoprecipitation and western blotting. (F. RAUSCHER and L. BENJAMIN, unpublished data). Thus, the fusion of EWS-WT1 is highly specific to the DSRCT phenotype and, like other fusions in the EWS/TLS family (Fig. 1) is predicted to encode a dominant oncogenic transcription factor.

Most important for this discussion is to consider the alteration to the biochemical functions of WT1. Figure 2 illustrates the wild-type WT1 and EWS1 genes as well as the resultant EWS–WT1 fusion protein. A model for how the fusion may function as an antagonist of wild-type WT1 is provided in Figure 3. To begin, since WT1 is a tumor suppressor gene, we must consider the consequences of both loss and gain of function. The most dramatic physical alteration of WT1 is replacement of its normal NH_2-terminus with the NTD of EWS. The glutamine-proline-rich NH_2-terminus of WT1 is capable of mediating both repression and activation of

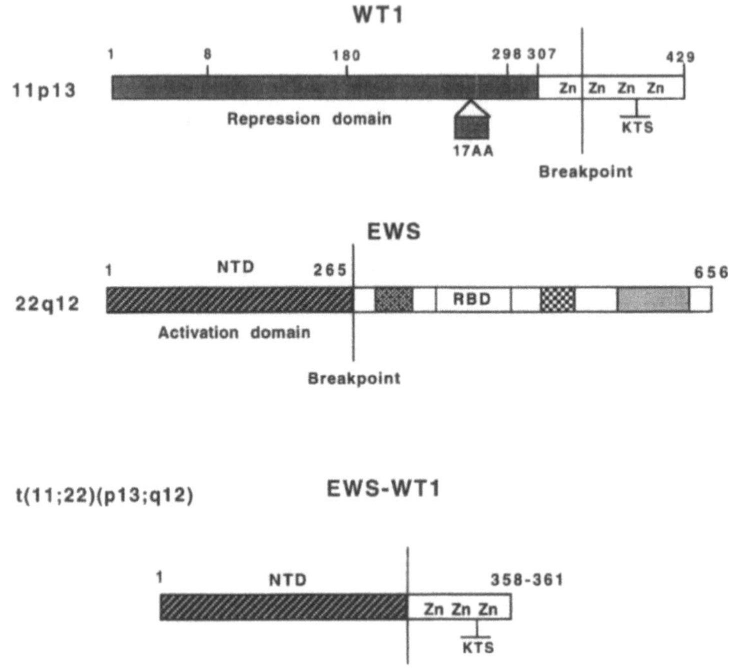

Fig. 2. Products of the t(11;22) translocation in DSRCT. The relative locations of the genomic breakpoints in each gene are shown. A fusion mRNA encodes the NTD of EWS1 fused in-frame to the zinc finger region of WT1

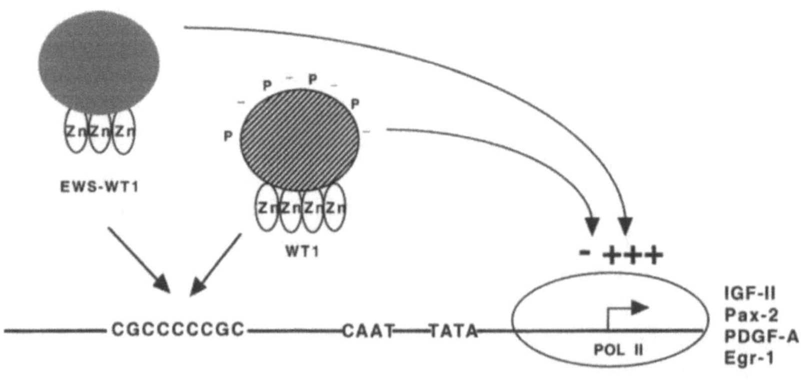

Fig. 3. A model to account for dominant oncogenesis mediated by the EWS-WT1 fusion protein. Fusion of EWS and WT1 genes creates a potent activator of transcription which displays a loosened sequence specificity for DNA recognition. A gain-of-function mutation in a tumor suppressor gene creates a dominant oncogene

transcription depending on cell type and binding site context within the target promoter (MADDEN et al. 1991). In addition, the WT1 NH_2-terminus mediates homo-oligomerization. It is likely that these functions are normally highly regulated via post-translational modification, presence of associated proteins (MAHESWARAN et al. 1993) and possibly editing of the RNA transcript (SHARMA et al. 1994). However, in DSRCT, the NH_2-terminus of WT1 is replaced by a potent transcriptional activation domain derived from EWS. When assayed under conditions where wild-type WT1 functions as a transcriptional repressor, EWS-WT1 functions as a potent activator of transcriptional (RAUSCHER et al. 1994). Thus, like other EWS fusions, the oncogenic potential of the EWS-WT1 fusion may involve activation of physiologically relevant target genes normally repressed by wild-type WT1.

The second alteration to a discrete functional domain of WT1 is loss of the NH_2-terminal zinc finger. Paradoxically, loss of zinc finger 1 appears to activate the DNA binding potential of the remaining three zinc fingers when assayed with the EGR-consensus oligonucleotide sequence (RAUSCHER et al. 1994). We have previously shown that finger 1 provides additional specificity for DNA recognition and destabilizes DNA binding activity of the wild-type, four finger WT1 protein at some DNA binding sites (MADDEN and RAUSCHER 1993; DRUMMOND et al. 1994). Furthermore, loss of finger 1 function (via site-directed mutagenesis or overt deletion) enhances DNA binding affinity. Thus, the results of our in vitro structure–function analysis of requirements for WT1 DNA binding parallel the effects of naturally occurring mutations in DSRCT. The loss of sequence specificity and/or gain in DNA binding affinity displayed by the EWS-WT1 fusion protein may expand the repertoire of downstream target genes normally subject to WT1-mediated transcriptional control. The fact that finger 2, 3, and 4 of WT1 most closely resemble the three finger EGR proteins, the normal target genes for EGR may also be deregulated by EWS-WT1. A parallel issue which must be addressed is the effect of loss of finger 1 on the still undefined RNA binding activities of WT1.

5 Summary and Future Directions

A variety of independent cytogenetic and biologic lines of research culminated in the analysis of the EWS1 and WT1 genes in desmoplastic round cell tumors and their refinement as a common lesion in this distinct tumor entity. From a purely genetic standpoint, a single allele of each gene involved in the translocation is inactivated as a result of the fusion. From what we know of the tumor suppressor capability of WT1, this somatic mutation may simply inactivate function and contribute to the malignant process via a gene dosage mechanism. This is entirely consistent with Wilms' tumor genetics in that when alterations in WT1 are found in sporadic tumors (~10% frequency) they are almost exclusively heterozygous in nature.

However, the biochemical analysis of the EWS-WT1 fusion argues against this simplistic interpretation. Clearly WT1 has suffered a gain-of-function alteration. The protein displays a somewhat relaxed recognition specificity for DNA binding. This is in contrast to all other EWS fusions which leave the DNA binding domain intact (Fig. 1). Moreover, the chimeric EWS-WT1 protein is converted from a repressor into an activator of transcription as a result of fusion to the EWS-NTD. Although there are relatively few physiologically relevant functional assays with which to compare WT1 and EWS-WT1 by analogy to other EWS fusions (Fig. 2) it is likely that fusion to the NTD is a major component of the oncogenic mechanism. In addition, it is quite likely that as yet ill-defined layers of regulation spanning from post-translational and transcriptional modifications to protein-protein interactions are disrupted as a result of loss of the WT1 NH_2-terminus. It is these areas that must be addressed in the coming years. Essentially the recognition that a common transcriptional effect domain (the NTD) is fused to a multitude of DNA binding domains thereby creating a set of oncogenes has created a remarkable laboratory for the study of transcription regulation cell growth. Study of the EWS/TLS family of fusion oncogenes should provide a major new insight into these issues in the future.

Acknowledgements. I am grateful to L. Benjamin, W. Fredericks, and F. Barr for participating in the studies summarized herein. F. Rauscher's laboratory is supported by grants CA10815, DK49210, GM54220, DAMD17-96-1-6141, ACS grant NP954, the Irving A. Hansen Memorial Foundation, the Mary A. Rumsey Memorial Foundation, and the Pew Scholars Program in the Biomedical Sciences.

References

Beckwith JB, Kiviat NB, Bonadio JF (1990) Nephrogenic rests, nebroblastomatosis and the pathogenesis of Wilms' tumor. Pediatr Pathol 10: 1–36

Biegel JA, Conard K, Brooks JJ (1993) Translocation (11;22)(p13;q12): Primary change in intra-abdominal desmoplastic small round cell tumor. Genes Chromosome Cancer 7: 119–121

Call KM, Glaser T, Ito CY, Buckler AJ, Pelletier J, Haber DA, Rose EA, Kral A, Yeger H, Lewis WH, Jones C, Housman DE (1990) Isolation and characterization of a zinc finger polypeptide gene at the human chromosome 11 Wilms' tumor locus. Cell 60: 509–520

Cook DM, Hinkes MT, Bernfield M, Rauscher III FJ (1996) Transcriptional activation of the Syndecan-1 promoter by the Wilms' tumor protein WT1. Oncogene 13: 1789–1799

Coppes MJ, Campbell CE, Williams BRG (1993) The role of WT1 in Wilms' tumorigenesis. FASEB J 7: 886–895

Crozat A, Aman P, Mandahl N, Ron D (1993) Fusion of CHOP to a novel RNA-binding protein in human myxoid liposarcoma. Nature 363: 640–644

de Alava E, Ladanyi M, Rosai J, Gerald WL (1995) Detection of chimeric transcripts in desmoplastic small round cell tumor and related developmental tumors by reverse transcriptase polymerase chain reaction. 147: 1584–1590

Delattre O, Zucman J, Plougastel B, Desmaze C, Melot T, Peter M, Kovar H, Joubert I, de Jong P, Rouleau G, Aurias A, Thomas G (1992) Gene fusion with an ETS DNA-binding domain caused by chromosome translocation in human tumors. Nature 359: 162–165

Delattre O, Zucman J, Melot T, Garau XS, Zucker J-M, Lenoir GM, Ambros PF, Sheer D, TurcCarel C, Triche TJ, Aurias A, Thomas G (1994) The Ewing family of tumors – a subgroup of small-round-cell tumors defined by specific chimeric transcripts. N Engl J Med 331: 294–299

Dey SR, Sukhatme VP, Roberts AB, Sporn MB, Rauscher FJ III, Kim S-J (1994) Repression of the transforming growth factor-beta 1 gene by the Wilms' tumor suppressor WT1 gene product. Mol Endocrinol 8: 595–602

Drummond IA, Madden SL, Rowher-Nutter P, Bell GI, Sukhatme VP, Rauscher III FJ (1992) Repression of the insulin-like growth factor-II gene by the Wilms' tumor suppressor WT1. Science 257: 674–678

Drummond IA, Rupprecht HD, Rohwer-Nutter P, Lopez-Guisa JM, Madden SL, Rauscher III FJ, Sukhatme VP (1994) DNA recognition by variants of the Wilms' tumor suppressor, WT1. Mol Cell Biol 14: 3800–3809

Gashler AL, Bonthron DT, Madden SL, Rauscher III FJ, Collins T, Sukhatme VP (1992) Human platelet-derived growth factor A chain is transcriptionally repressed by the Wilms' tumor suppressor WT1. Proc Natl Acad Sci USA 89: 10984–10988

Gerald WL, Miller HK, Battifora H, Miettinen M, Silva EG, Rosai J (1991) Intra-abdominal desmoplastic small round-cell tumor. Report of 19 cases of a distinctive type of high-grade polyphenotypic malignancy affecting young individuals. Am J Surg Pathol 15: 499–513

Goodyer P, Dehbi M, Torban E, Bruening W, Pelletier J (1995) Repression of the retinoic acid receptor-alpha gene by the Wilms' tumor suppressor gene product, wt1. Oncogene 10: 1125–1129

Haber DA, Housman DE (1992) The genetics of Wilms' tumor. Adv Cancer Res 59: 41–68

Haber DA, Buckler AJ, Glaser T, Call KM, Pelletier J, Sohn RL, Douglass EC, Housman DE (1990) An internal deletion within in 11p13 zinc finger gene contributes to the development of Wilms' tumor. Cell 61: 1257–1269

Haber DA, Sohn RL, Buckler AJ, Pelletier J, Call KM, Housman DE (1991) Alternative splicing and genomic structure of the Wilms' tumor gene WT1. Proc Natl Acad Sci USA 88: 9618–9622

Haber DA, Timmer HT, Pelletier J, Sharp PA, Housman DE (1992) A dominant mutation in Wilms' tumor gene WT1 cooperatives with the viral oncogene EIA in transformation of primary kidney cells. Proc Natl Acad Sci USA 89: 6010–6014

Jeon IS, Davis JN, Braun BS, Sublett JE, Roussel MF, Denny CT, Shapiro DN (1995) A variant Ewing's sarcoma translocation (7;22) fuses the EWS gene to the ETS gene ETV1. Oncogene 10: 1229–1234

Kaneko Y, Yoshida K, Handa M, Toyoda Y, Nishikira H, Tanaka Y, Sasaki Y, Ishida S, Higashino F, Fujinaga K (1996) Fusion of an *ETS*-family gene, *EIAF*, to EWS by t(17;22)(q12,q12) chromosome translocation in an undifferentiated sarcoma of infancy. Genes Chrom Cancer 15: 115–121

Kreidberg JA, Sariola H, Loring JM, Maeda M, Pelletier J, Housman D, Jaenisch R (1993) WT-1 is required for early kidney development. Cell 74: 679–691

Labelle Y, Zucman J, Stenman G, Kindblom L-G, Knight J, Turc-Carel C, Dockhorn-Dworniczak B, Mandahl N, Desmaze C, Peter M, Aurias A, Delattre O, Thomas G (1995) Oncogenic conversion of a novel orphan nuclear receptor by chromosome translocation. Hum Mol Gen 4: 2219–2226

Ladanyi M, Gerald W (1994) Fusion of the EWS and WT1 genes in the desmoplastic small round cell tumor. Cancer Res 54: 2837–2840

Larsson SH, Charlieu J-P, Miyagawa K, Engelkamp D, Rassoulzadegan M, Ross A, Cuzin F, van Heyningen V, Hastie ND (1995) Subnuclear localization of WT1 in splicing or transcription factor domains is regulated by alternative splicing Cell 81: 391–401

Madden SL, Rauscher III FJ (1993) Positive and negative regulation of transcription and cell growth mediated by the EGR family of zinc finger gene products. In: Zinc-finger proteins in oncogenesis: DNA-binding and gene regulation. Ann NY Acad Sci 684: 75–84

Madden SL, Cook DM, Morris JF, Gashler A, Sukhatme VP, Rauscher III FJ (1991) Transcriptional repression mediated by the WT1 Wilms' tumor gene product. Science 253: 1550–1553

Maheswaran S, Park S, Bernard A, Morris JF, Rauscher III FJ, Hill DE, Haber DA (1993) Interaction between the p53 and Wilms' tumor (WT1) gene products: physical association and functional cooperation. Proc Natl Acad Sci USA 90: 5100–5104

Ordonez NG, Zirkin R, Bloom RE (1989) Malignant small-cell epithelial tumor of the peritoneum coexpressing mesenchymal-type intermediate filaments. Am J Surg Pathol 13: 413–421

Ordonez NG, El-Naggar AK, Ro JY, Silva EG, Mackay B (1993) Intra-abdominal desmoplastic small cell tumor: a light microscopic, immunocytochemical, ultrastructural, and flow cytometric study. Hum Pathol 24: 850–865

Pelletier J, Bruening W, Kashtan CE, Mauer SM, Manivel JC, Striegel JE, Houghton DC, Junien C, Habib R, Fouser L, Fine RN, Silverman RL, Haber DA, Housman D (1991) Germline mutations in the Wilms' tumor suppressor gene are associated with abnormal urogenital development in Denys-Drash syndrome. Cell 67: 437–447

Pritchard-Jones K, Fleming S, Davidson D, Bickmore W, Porteous D, Gosden C, Bard J, Buckler A,

Pelletier J, Housman D, van Heyningen V, Hastie N (1990) The candidate Wilms' tumor gene is involved in genitourinary development. Nature 346: 194–197
Rabbits TH, Forster A, Larson R, Nathan P (1993) Fusion of the dominant negative transcription regulator CHOP with a novel gene FUS by translocation t(12;16) in malignant liposarcoma. Nature Genet 4: 175–180
Rauscher III FJ, Morris JF, Tournay OE, Cook DM, Curran T (1990) Binding of the Wilms' tumor locus zinc finger protein in the EGR-1 consensus sequence. Science 250: 1259–1262
Rauscher III FJ, Benjamin LE, Fredericks WJ, Morris JF (1994) Novel oncogenic mutations in the WT1 Wilms' tumor suppressor gene. A recurrent t(11;22) fuses the Ewings' Sarcoma gene, EWS1 to WT1 in desmoplastic small round cell tumor. Gold Spring Harbor Symposium on Quantitative Biology. Cold Spring Harbor 59: 137–146
Rodriguez E, Sreekantaiah C, Gerald W, Reuter VE, Motzer RJ, Chaganti RSK (1993) A recurring translocation t(11;22)(p12;q11.2) characterizes intra-abdominal desmoplastic small round-cell tumors. Cancer Genet Cytogenet 69: 17–21
Ryan G, Steele-Perkins V, Morris JM, Rauscher III FJ, Dressler GR (1995) Repression of Pax-2 by WT1 during normal kidney development. Development 129: 867–875
Sawyer JR, Tryka AF, Lewis JM (1992) A novel reciprocal chromosome translocation t(11;22)(p13;q12) in an intraabdominal desmoplastic small round-cell tumor. Am J Surg Pathol 16: 411–416
Saxen L (1987) Organogenesis of the kidney. Cambridge University Press, Cambridge
Sharma PM, Bowman M, Madden SL, Rauscher III FJ, Sukumar S (1994) RNA editing in the Wilms' tumor susceptibility gene, WT1. Genes Dev 8: 720–731
Shen WP, Towne B, Zadeh TM (1992) Cytogenetic abnormalities in an intra-abdominal desmoplastic small cell tumor. Cancer Genet Cytogenet 64: 189–191
Sorenson PHB, Lessnick SL, Lopez-Terrada D, Liu XF, Triche TJ, Denny CT (1994) A second Ewing's sarcoma translocation t(21;22) fuses the EWS gene to another ETS-family transcription factor, ERG. Nature Genet 6: 146–151
Sukhatme VP (1990) Early transcriptional events in cell growth: the Egr family. J Am Soc Nephrol 1: 859–866
Urano F, Umezawa A, Hong W, Kikuchi H, Hata J-i (1996) A novel chimera gene between *EWS* and *EIA-F*, encoding the adenovirus E1A enhancer-binding protein, in extraosseous Ewing's Sarcoma. 219: 608–61
Wang ZY, Madden SL, Deuel TF, Rauscher III FJ (1992) The human platelet-derived growth factor A-chain (PDGF-A) gene is a target for repression by the WT1 Wilms' tumor protein. J Biol Chem 267: 21999–22002
Werner H, Rauscher III FJ, Sukhatme VP, Drummond IA, Roberts Jr CT, LeRoith D (1994) Transcriptional repression of the insulin-like growth factor I receptor (IGF-I-R) gene by the tumor suppressor WT1 involves binding to sequences upstream and downstream of the IGF-I-R transcription start site. J Biol Chem 269: 12577–12582
Zucman J, Delattre O, Desmaze C, Epstein A, Stenman G, Speleman F, Fletchers CDM, Aurias A, Thomas G (1993) EWS and ATF-1 gene fusion induced by t(12;22) translocation in malignant melanoma of soft parts. Nature Genet 4: 341–345

Subject Index

Current Topics in Microbiology and Immunology

Volumes published since 1989 (and still available)

Vol. 180: **Sansonetti, P. J. (Ed.):** Pathogenesis of Shigellosis. 1992. 15 figs. X, 143 pp. ISBN 3-540-55058-5

Vol. 181: **Russell, Stephen W.; Gordon, Siamon (Eds.):** Macrophage Biology and Activation. 1992. 42 figs. IX, 299 pp. ISBN 3-540-55293-6

Vol. 182: **Potter, Michael; Melchers, Fritz (Eds.):** Mechanisms in B-Cell Neoplasia. 1992. 188 figs. XX, 499 pp. ISBN 3-540-55658-3

Vol. 183: **Dimmock, Nigel J.:** Neutralization of Animal Viruses. 1993. 10 figs. VII, 149 pp. ISBN 3-540-56030-0

Vol. 184: **Dunon, Dominique; Mackay, Charles R.; Imhof, Beat A. (Eds.):** Adhesion in Leukocyte Homing and Differentiation. 1993. 37 figs. IX, 260 pp. ISBN 3-540-56756-9

Vol. 185: **Ramig, Robert F. (Ed.):** Rotaviruses. 1994. 37 figs. X, 380 pp. ISBN 3-540-56761-5

Vol. 186: **zur Hausen, Harald (Ed.):** Human Pathogenic Papillomaviruses. 1994. 37 figs. XIII, 274 pp. ISBN 3-540-57193-0

Vol. 187: **Rupprecht, Charles E.; Dietzschold, Bernhard; Koprowski, Hilary (Eds.):** Lyssaviruses. 1994. 50 figs. IX, 352 pp. ISBN 3-540-57194-9

Vol. 188: **Letvin, Norman L.; Desrosiers, Ronald C. (Eds.):** Simian Immunodeficiency Virus. 1994. 37 figs. X, 240 pp. ISBN 3-540-57274-0

Vol. 189: **Oldstone, Michael B. A. (Ed.):** Cytotoxic T-Lymphocytes in Human Viral and Malaria Infections. 1994. 37 figs. IX, 210 pp. ISBN 3-540-57259-7

Vol. 190: **Koprowski, Hilary; Lipkin, W. Ian (Eds.):** Borna Disease. 1995. 33 figs. IX, 134 pp. ISBN 3-540-57388-7

Vol. 191: **ter Meulen, Volker; Billeter, Martin A. (Eds.):** Measles Virus. 1995. 23 figs. IX, 196 pp. ISBN 3-540-57389-5

Vol. 192: **Dangl, Jeffrey L. (Ed.):** Bacterial Pathogenesis of Plants and Animals. 1994. 41 figs. IX, 343 pp. ISBN 3-540-57391-7

Vol. 193: **Chen, Irvin S. Y.; Koprowski, Hilary; Srinivasan, Alagarsamy; Vogt, Peter K. (Eds.):** Transacting Functions of Human Retroviruses. 1995. 49 figs. IX, 240 pp. ISBN 3-540-57901-X

Vol. 194: **Potter, Michael; Melchers, Fritz (Eds.):** Mechanisms in B-cell Neoplasia. 1995. 152 figs. XXV, 458 pp. ISBN 3-540-58447-1

Vol. 195: **Montecucco, Cesare (Ed.):** Clostridial Neurotoxins. 1995. 28 figs. XI., 278 pp. ISBN 3-540-58452-8

Vol. 196: **Koprowski, Hilary; Maeda, Hiroshi (Eds.):** The Role of Nitric Oxide in Physiology and Pathophysiology. 1995. 21 figs. IX, 90 pp. ISBN 3-540-58214-2

Vol. 197: **Meyer, Peter (Ed.):** Gene Silencing in Higher Plants and Related Phenomena in Other Eukaryotes. 1995. 17 figs. IX, 232 pp. ISBN 3-540-58236-3

Vol. 198: **Griffiths, Gillian M.; Tschopp, Jürg (Eds.):** Pathways for Cytolysis. 1995. 45 figs. IX, 224 pp. ISBN 3-540-58725-X

Vol. 199/I: **Doerfler, Walter; Böhm, Petra (Eds.):** The Molecular Repertoire of Adenoviruses I. 1995. 51 figs. XIII, 280 pp. ISBN 3-540-58828-0

Vol. 199/II: **Doerfler, Walter; Böhm, Petra (Eds.):** The Molecular Repertoire of Adenoviruses II. 1995. 36 figs. XIII, 278 pp. ISBN 3-540-58829-9

Vol. 199/III: **Doerfler, Walter; Böhm, Petra (Eds.):** The Molecular Repertoire of Adenoviruses III. 1995. 51 figs. XIII, 310 pp. ISBN 3-540-58987-2

Vol. 200: **Kroemer, Guido; Martinez-A., Carlos (Eds.):** Apoptosis in Immunology. 1995. 14 figs. XI, 242 pp. ISBN 3-540-58756-X

Vol. 201: **Kosco-Vilbois, Marie H. (Ed.):** An Antigen Depository of the Immune System: Follicular Dendritic Cells. 1995. 39 figs. IX, 209 pp. ISBN 3-540-59013-7

Vol. 202: **Oldstone, Michael B. A.; Vitković, Ljubiša (Eds.):** HIV and Dementia. 1995. 40 figs. XIII, 279 pp. ISBN 3-540-59117-6

Vol. 203: **Sarnow, Peter (Ed.):** Cap-Independent Translation. 1995. 31 figs. XI, 183 pp. ISBN 3-540-59121-4

Vol. 204: **Saedler, Heinz; Gierl, Alfons (Eds.):** Transposable Elements. 1995. 42 figs. IX, 234 pp. ISBN 3-540-59342-X

Vol. 205: **Littman, Dan R. (Ed.):** The CD4 Molecule. 1995. 29 figs. XIII, 182 pp. ISBN 3-540-59344-6

Vol. 206: **Chisari, Francis V.; Oldstone, Michael B. A. (Eds.):** Transgenic Models of Human Viral and Immunological Disease. 1995. 53 figs. XI, 345 pp. ISBN 3-540-59341-1

Vol. 207: **Prusiner, Stanley B. (Ed.):** Prions Prions Prions. 1995. 42 figs. VII, 163 pp. ISBN 3-540-59343-8

Vol. 208: **Farnham, Peggy J. (Ed.):** Transcriptional Control of Cell Growth. 1995. 17 figs. IX, 141 pp. ISBN 3-540-60113-9

Vol. 209: **Miller, Virginia L. (Ed.):** Bacterial Invasiveness. 1996. 16 figs. IX, 115 pp. ISBN 3-540-60065-5

Vol. 210: **Potter, Michael; Rose, Noel R. (Eds.):** Immunology of Silicones. 1996. 136 figs. XX, 430 pp. ISBN 3-540-60272-0

Vol. 211: **Wolff, Linda; Perkins, Archibald S. (Eds.):** Molecular Aspects of Myeloid Stem Cell Development. 1996. 98 figs. XIV, 298 pp. ISBN 3-540-60414-6

Vol. 212: **Vainio, Olli; Imhof, Beat A. (Eds.):** Immunology and Developmental Biology of the Chicken. 1996. 43 figs. IX, 281 pp. ISBN 3-540-60585-1

Vol. 213/I: **Günthert, Ursula; Birchmeier, Walter (Eds.):** Attempts to Understand Metastasis Formation I. 1996. 35 figs. XV, 293 pp. ISBN 3-540-60680-7

Vol. 213/II: **Günthert, Ursula; Birchmeier, Walter (Eds.):** Attempts to Understand Metastasis Formation II. 1996. 33 figs. XV, 288 pp. ISBN 3-540-60681-5

Vol. 213/III: **Günthert, Ursula; Schlag, Peter M.; Birchmeier, Walter (Eds.):** Attempts to Understand Metastasis Formation III. 1996. 14 figs. XV, 262 pp. ISBN 3-540-60682-3

Vol. 214: **Kräusslich, Hans-Georg (Ed.):** Morphogenesis and Maturation of Retroviruses. 1996. 34 figs. XI, 344 pp. ISBN 3-540-60928-8

Vol. 215: **Shinnick, Thomas M. (Ed.):** Tuberculosis. 1996. 46 figs. XI, 307 pp. ISBN 3-540-60985-7

Vol. 216: **Rietschel, Ernst Th.; Wagner, Hermann (Eds.):** Pathology of Septic Shock. 1996. 34 figs. X, 321 pp. ISBN 3-540-61026-X

Vol. 217: **Jessberger, Rolf; Lieber, Michael R. (Eds.):** Molecular Analysis of DNA Rearrangements in the Immune System. 1996. 43 figs. IX, 224 pp. ISBN 3-540-61037-5

Vol. 218: **Berns, Kenneth I.; Giraud, Catherine (Eds.):** Adeno-Associated Virus (AAV) Vectors in Gene Therapy. 1996. 38 figs. IX,173 pp. ISBN 3-540-61076-6

Vol. 219: **Gross, Uwe (Ed.):** Toxoplasma gondii. 1996. 31 figs. XI, 274 pp. ISBN 3-540-61300-5

MIX
Papier aus verantwortungsvollen Quellen
Paper from responsible sources
FSC® C105338

If you have any concerns about our products,
you can contact us on
ProductSafety@springernature.com

In case Publisher is established outside the EU,
the EU authorized representative is:
Springer Nature Customer Service Center GmbH
Europaplatz 3, 69115 Heidelberg, Germany

Printed by Libri Plureos GmbH
in Hamburg, Germany